T0134571

Studies in Computational Intelligence

Volume 559

Series editor

Janusz Kacprzyk, Polish Academy of Sciences, Warsaw, Poland
e-mail: kacprzyk@ibspan.waw.pl

For further volumes:
http://www.springer.com/series/7092

About this Series

The series "Studies in Computational Intelligence" (SCI) publishes new developments and advances in the various areas of computational intelligence—quickly and with a high quality. The intent is to cover the theory, applications, and design methods of computational intelligence, as embedded in the fields of engineering, computer science, physics and life sciences, as well as the methodologies behind them. The series contains monographs, lecture notes and edited volumes in computational intelligence spanning the areas of neural networks, connectionist systems, genetic algorithms, evolutionary computation, artificial intelligence, cellular automata, self-organizing systems, soft computing, fuzzy systems, and hybrid intelligent systems. Of particular value to both the contributors and the readership are the short publication timeframe and the world-wide distribution, which enable both wide and rapid dissemination of research output.

Zdzisław S. Hippe · Juliusz L. Kulikowski
Teresa Mroczek · Jerzy Wtorek
Editors

Issues and Challenges in Artificial Intelligence

Editors
Zdzisław S. Hippe
Teresa Mroczek
University of Information Technology
and Management
Rzeszów
Poland

Jerzy Wtorek
Gdańsk University of Technology
Gdańsk-Wrzeszcz
Poland

Juliusz L. Kulikowski
Institute of Biocybernetics and Biomedical
Engineering
Polish Academy of Sciences
Warszawa
Poland

ISSN 1860-949X ISSN 1860-9503 (electronic)
ISBN 978-3-319-36143-7 ISBN 978-3-319-06883-1 (eBook)
DOI 10.1007/978-3-319-06883-1
Springer Cham Heidelberg New York Dordrecht London

Softcover reprint of the hardcover 1st edition 2014

Printed on acid-free paper

Springer is part of Springer Science+Business Media (www.springer.com)

Preface

Together with the increase in computer technology development the importance of human–computer system interaction problems is increasing due to the growing expectations of users on general computer systems' capabilities in human work and life facilitation. Users expect a system that is not a passive tool in human hands, but rather an active partner equipped with a sort of artificial intelligence, having access to large information resources, being able to adapt its behavior to human requirements, and to collaborate with human users. Achieving these expectations is possible through interaction with technology-based systems (e.g., computers, embedded computer devices) through interactive modalities (e.g., video, voice, touching, writing, gesture, facial expressions, and many others). Thus, the computational methods of Artificial Intelligence are inherent tools utilized in this research area. It is an idea of the book titled *Issues and Challenges in Artificial Intelligence* to collect examples of such attempts. Its preparation was made possible thanks to the papers sent in by our colleagues, working as we do, in the area of Human Interaction Systems. We appreciate these contributions very much.

The contents of the book were divided into the following parts: Part I. Detection, Recognition and Reasoning; Part II. Data Modeling, Acquisition and Mining; and Part III. Optimization.

Part I, consisting of six chapters, is devoted to detection, recognition, and reasoning in different circumstances and applications. H. M. Nguyen et al. evaluate a hybrid approach to reconstruct a wide range of 3D objects from photographs. A. Lipnickas et al. tune parameters of the RANSAC method and determine the number of planes in 3D data by means of indices describing validity of clusters. In another chapter, these authors again adopt the RANSAC method for segmentation of flat areas and detection of planes. A new approach in automatic speech recognition is presented in the chapter by R. Amami and coworkers. It is based on a non-conventional utilization of Real Adaboost algorithm in combination with Support Vector Machines. A. Smiti and Z. Elouedi recognize the importance of the clustering exploitation in competence computing for Case-Based Reasoning systems. B. Sniezynski et al. discuss a methodology for application of Logic Plausible Reasoning formalism in a creation of specific knowledge.

Problems associated with data modeling, acquisition, and mining are presented in papers collected in Part II. A comparison of two approaches used for intelligent planning of complex chemical synthesis is presented by Z. S. Hippe.

In effect, the results of experiments enhancing the matrix model of constitutional chemistry by machine learning algorithms put a new quality into the worldwide known **D-U** model. The chapter by P. G. Clark and coworkers deals with generalized probabilistic approximations applicable in mining inconsistent data. J. L. Kulikowski presents a computer-aided assessment of complex effects of decisions and shows that finding trade-offs between costs and profits is possible due to adequately chosen algebraic tools. A new concept of a distributed system for data acquisition, preprocessing, and subsequent passing by modern mobile devices is discussed by P. Czarnul. An exemplary implementation of the concept on modern Phonegap platform is also provided in this chapter.

Part III contains four chapters. J. Balicki et al. consider a genetic scheduler applied for optimization of a bottleneck computer workload and costs. Genetic programming is applied for finding the Pareto solutions by applying an immunological procedure. The authors conclude that a computer program as a chromosome gives the possibility to represent specific knowledge of the considered problem in a more intelligent way than the data structure. C. Barbulescu and S. Kilyeni examine, in their chapter, the particle swarm optimization algorithm applied to study the power flow in complex power systems. T. Potuzak, in his chapter, describes time requirements of genetic algorithm optimization for road traffic division when using a distributed version of the algorithm. In turn, A. P. Rotshtein and H. B. Rakytynska discuss optimal design of rule-based system by means of solving fuzzy relational equations. The proposed approach leads to achieving an optimal accuracy-complexity trade-off as a result of the total number of decision classes optimization.

We hope that this book will find a wide audience of readers and that they find it an interesting one.

Zdzisław S. Hippe
Juliusz L. Kulikowski
Teresa Mroczek
Jerzy Wtorek

Contents

Part I
Detection, Recognition and Reasoning

A Robust System for High-Quality Reconstruction of 3D Objects from Photographs

H. M. Nguyen, B. C. Wünsche, P. Delmas, C. Lutteroth and W. van der Mark

Abstract Image-based modeling is rapidly increasing in popularity since cameras are very affordable, widely available, and have a wide image acquisition range suitable for objects of vastly different size. In this chapter we describe a novel image-based modeling system, which produces high-quality 3D content automatically from a collection of unconstrained and uncalibrated 2D images. The system estimates camera parameters and a 3D scene geometry using *Structure-from-Motion (SfM)* and *Bundle Adjustment* techniques. The point cloud density of 3D scene components is enhanced by exploiting silhouette information of the scene. This hybrid approach dramatically improves the reconstruction of objects with few visual features. A high quality texture is created by parameterizing the reconstructed objects using a segmentation and charting approach, which also works for objects which are not homeomorphic to a sphere. The resulting parameter space contains one chart for each surface segment. A texture map is created by back projecting the best fitting input images onto each surface segment, and smoothly fusing them together over the corresponding chart by using graph-cut techniques. Our evaluation shows that our system is capable of reconstructing a wide range of objects in both indoor and outdoor environments.

H. M. Nguyen · B. C. Wünsche (✉) · P. Delmas · C. Lutteroth · W. van der Mark
Department of Computer Science, University of Auckland, Auckland, New Zealand
e-mail: burkhard@cs.auckland.ac.nz

H. M. Nguyen
e-mail: justin.nguyen@auckland.ac.nz

P. Delmas
e-mail: p.delmas@cs.auckland.ac.nz

C. Lutteroth
e-mail: lutteroth@cs.auckland.ac.nz

W. van der Mark
e-mail: w.vandermark@auckland.ac.nz

Z. S. Hippe et al. (eds.), *Issues and Challenges in Artificial Intelligence*,
Studies in Computational Intelligence 559, DOI: 10.1007/978-3-319-06883-1_1,

1 Introduction

A key task in mobile robotics is the exploration and mapping of an unknown environment using the robot's sensors. SLAM algorithms can create a map in real time using different sensors. While the resulting map is suitable for navigation, it usually does not contain a high quality reconstruction of the surrounding 3D scene, e.g., for use in virtual environments, simulations, and urban design.

High quality reconstructions can be achieved using image input and conventional modeling systems such as Maya, Lightwave, 3D Max or Blender. However, the process is time-consuming, requires artistic skills, and involves considerable training and experience in order to master the modeling software. The introduction of specialized hardware has simplified the creation of models from real physical objects. Laser scanners can create highly accurate 3D models, but are expensive and have a limited range and resolution. RGBD sensors, such as the Kinect, have been successfully used for creating large scale reconstructions. In 2011 the KinectFusion algorithm was presented, which uses the Kinect depth data to reconstruct a 3D scene using the Kinect sensor like a handheld laser scanner (Newcombe et al. 2011). Since then a wide variety of new applications have been proposed such as complete 3D mappings of environments (Henry et al. 2012). The Kinect is very affordable, but has a very limited operating range (0.8–3.5 m), a limited resolution and field-of-view, and it is sensitive to environmental conditions (Oliver et al. 2012). Reconstruction 3D scenes from optical sensor data has considerable advantages such as the low price of cameras, the ability to capture objects of vastly different size, and the ability to capture highly detailed color and texture information. Furthermore optical sensors are very light weight and have a low energy consumption, which makes them ideal for mobile robots, such as small Unmanned Aerial Vehicles (UAVs).

This chapter proposes a novel system that employs a hybrid multi-view image-based modeling approach coupled with a surface parameterization technique as well as surface and texture reconstruction for automatically creating a high quality reconstruction of 3D objects using uncalibrated and unconstrained images acquired using consumer-level cameras. In contrast to previous works we combine both correspondence-based and silhouette-base reconstruction techniques, which improves reconstruction results for featureless objects and objects with concave regions. These classes of objects often pose great difficulty for algorithms using only a single approach. As the result, our solution is able to produce satisfactory results with higher resolution for a much larger class of objects.

The system performs 3D reconstruction using the following steps:

1. camera parameter estimation and scene geometry generation
2. increase the density of the obtained point cloud by exploiting object's silhouette information
3. 3D surface reconstruction
4. surface parameterisation and texture reconstruction.

The remainder of this chapter is structured as follows. In Sect. 2, we review related work in the field of image-based modeling. Section 3 presents the design of our solution. Results are discussed in Sect. 4. In Sect. 5 we conclude the chapter and discuss directions for future research.

2 Previous Works

3D image-based reconstruction algorithms can be classified and categorized based on the visual cues used to perform reconstruction, e.g., silhouettes, texture, shading or correspondence. Amongst them, shape-from-silhouette and shape-from-correspondence have proven to be the most well-known and successful visual cues. Classes of reconstruction methods exploiting these visual cues can offer a high degree of robustness due to their invariance to illumination changes (Hernandez et al. 2008).

Shape-from-silhouette algorithms obtain the 3D structure of an object by establishing an approximate maximal surface, known as the visual hull, which progressively encloses the actual object. Shape from silhouette-based methods can produce surprisingly good results with a relatively small number of views, but have problems with complex object geometries, such as concave regions (Grauman et al. 2003; Matusik et al. 2000; Nguyen et al. 2011). Most techniques extract silhouette contours (Baumgart 1974) and then derive a 3D geometry from them, e.g. by computing the intersection of silhouette cones (Martin et al. 1983). Efficiency can be improved by using an octree representation of the visual hull (Chien et al. 1984). Grauman et al. (2003) use a Bayesian approach to compensate for errors introduced as the result of false segmentation.

The literature in image-based modelling describes several complete systems, but only for a limited range of applications. Früh and Zakhor (2003) create textured 3D models of an entire city by using a combination of aerial imagery, ground color, and LIDAR scans, which makes the technique unpractical for consumer applications. Xiao et al. (2008) presented a semi-automatic image-based approach to reconstruct 3D façade models from a sequence of photographs. Quan et al. (2006) present a technique for modeling plants. The algorithm requires manual interaction and makes assumptions about the geometry of the reconstructed object (e.g. leaves).

3 Design

3.1 Algorithm Overview

In order to recover the scene geometry, our system automatically detects and extracts points of interest such as corners (edges with gradients in multiple directions) in the input images. The points are matched across views and changes of their relative position across multiple images are used to estimate camera

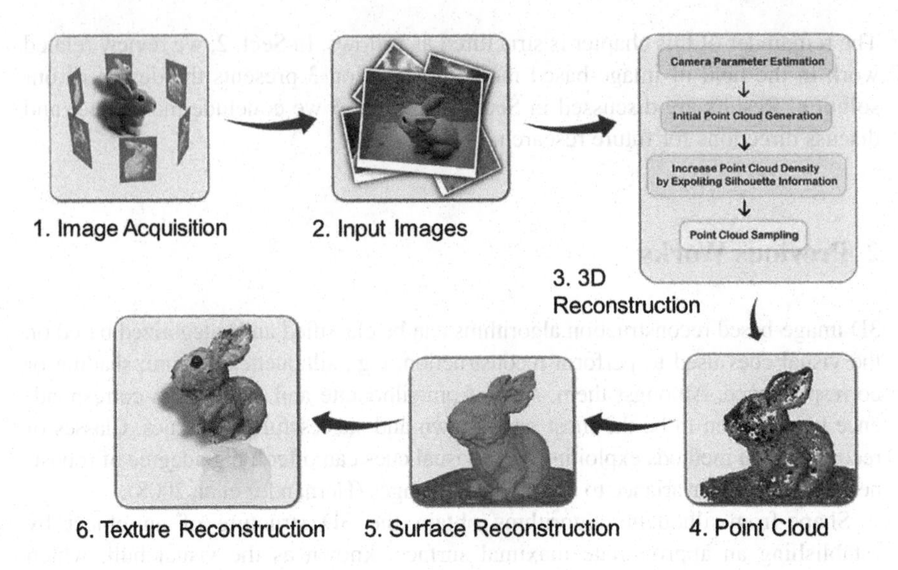

Fig. 1 Stages of the reconstruction process

parameters and 3D coordinates of the matched points using a *Structure from Motion* technique. The method requires that input images partially overlap.

Feature matching is achieved using an incremental approach starting with a pair or images having a large number of matches, but also a large baseline. This is to ensure that the 3D coordinates of observed points are well-conditioned. The remaining images are added one at a time ordered by the number of matches (Cheng et al. 2011; Snavely et al. 2006). The *Bundle Adjustment* technique is subsequently applied to refine and improve the solution.

The density of the obtained scene geometry is enhanced by exploiting the silhouette information in the input images. The end result of this stage is a dense point cloud of the scene to be reconstructed. A 3D surface mesh is obtained by interpolating the 3D point cloud. The surface is then parameterized and a texture map is obtained by back projecting the input images and fusing them together using graph-cut techniques. Figure 1 depicts several stages of the reconstruction process.

3.2 Camera Parameter Estimation

The objective of this stage is to recover the intrinsic and extrinsic parameters of each view. This is accomplished in two steps: First, salient features are extracted and matched across views. Second, the camera parameters are estimated using *Structure-from-Motion* and *Bundle Adjustment* techniques. In our system we use the SIFT feature detector (Lowe 1999, 2004).

Once features have been detected and extracted from the input images, they are matched in order to find pairwise correspondences between them. This is achieved by using a distance metric to compute the similarity of each feature of a candidate image with features of another image. A small distance signifies that the two key points are close and thus similar. However, a small distance does not necessarily mean that the points represent the same feature. For instance, the corners of windows of a building look similar regardless of whether two photos show the same or different parts of the building. In order to accurately match a key point in the candidate image, we identify the closest and second closest key point in the reference image using a nearest neighbor search strategy. If their ratio is below a given threshold, the key point and the closest matched key point are accepted as correspondences, otherwise the match is rejected (Lowe 1999, 2004).

At this stage, we have a set of potentially matching image pairs, and for each pair, a set of individual feature correspondences. For each pair of matching images, we compute the Fundamental Matrix using the RANSAC algorithm. Erroneous matches are eliminated by enforcing epipolar constraints. Scene geometry and the motion information of the camera are estimated using the *Structure-from-Motion* technique (Cheng et al. 2011; Snavely et al. 2006; Szeliski 2006), and are further refined using *Bundle Adjustment*.

3.3 Scene Geometry Enhancement

At this stage, we have successfully acquired both the camera parameters and the scene geometry. Due to the sparseness of the scene geometry, the surface and texture reconstruction frequently produce artefacts. Most previous works approached this problem by constraining the permissible object types or requiring manual hints for the reconstruction process. However, these requirements breach our goal of creating an easy-to-use system capable of reconstructing any type of object where shape and texture properties are correctly captured by the input photos.

We improve the reconstruction results by exploiting the silhouette information to further enrich the density of the point cloud: First, the silhouette information in each image is extracted using the *Marching Squares* algorithm (Lorensen et al. 1995), which produces a sequence of all contour pixels. To construct a visual hull representation of the scene using an immense silhouette contour point set will inevitably increase computational costs. In order to avoid this, the silhouette contour data is converted into a 2D mesh using a *Delaunay triangulation*, and the mesh is simplified using a mesh decimation algorithm (Melax 1998). This effectively reduces the number of silhouette contour points. A point cloud representing the visual hull of the scene is obtained using a technique presented by Matusik et al. (2000).

3.4 Surface Reconstruction

At this stage we have successfully obtained a quasi-dense 3D point cloud, which in the next step needs to be approximated by a smooth closed surface (without holes) that represents the underlying 3D model from which the point cloud was obtained. We tested several surface reconstruction techniques including the power crust algorithm (Amenta et al. 2001), α-shapes (Edelsbrunner 1995), and the ball-pivoting algorithm (Bernardini et al. 1999). We decided to employ the *Poisson Surface Reconstruction* algorithm (Kazhdan et al. 2006), since it produces a closed surface and works well for noisy data. In contrast to many other implicit surface fitting methods, which often segment the data into regions for local fitting and then combine these local approximations using blending functions, Poisson surface reconstruction processes all the sample points at once, without resorting to spatial segmentation or blending (Kazhdan et al. 2006).

3.5 Texture Reconstruction

A high-resolution texture for the reconstructed 3D object is obtained by parameterizing the 2D mesh and computing a texture map.

(a) *Surface Parameterization*: We tested surface parameterization algorithms provided by existing libraries and tools, such as Blender. We found that they required manual hints, only worked for objects homeomorphic to a sphere, or created a surface parameterization using many disconnected patches. The latter result is undesirable since it creates visible seams in the reconstructed texture, and since it makes post-processing steps, such as mesh reduction, more difficult.

In order to use the resulting 3D models in a large variety of applications and professional production pipelines, we need a texture map which consists of a small number of patches, which ideally correspond to geometric features (which can be maintained in a post-processing step such as mesh reduction). The Feature-based Surface Parameterization technique by Zhang et al. (2005) fulfils these criteria. The algorithm consists of three stages:

1. **Genus reduction**: In order to identify non-zero genus surfaces, a surface-based Reeb graph (Reeb 1946) induced by the average geodesic distance (Hilaga et al. 2001) is constructed. Cycles in the graph signify the existence of handles/holes in the surface, i.e., the surface is not homeomorphic to a sphere. Examples are donut and teacup shaped objects. The genus of the surface is reduced by cutting the surface along the cycles of the graph. The process is repeated until there are no more cycles.
2. **Feature identification**: Tips of surface protrusions are identified as leaves of the Reeb graph. The features are separated from the rest of the surface by constructing a closed curve.
3. **Patch creation**: The previous two steps segment the surface into patches which are homeomorphic to a disk. Patches are "unwrapped" using discrete conformal mappings (Eck et al. 1995). The algorithm first positions the texture coordinates of the boundary vertices, and then finds the texture coordinates of the interior vertices by solving a closed form system. Distortions are reduced by using a post-processing step, which optimizes the position of interior vertices' texture

Fig. 2 The Rooster model: the segmented 3D model and the corresponding texture atlas (*left*) and the reconstructed texture obtained by projecting and fusing input photographs (*right*)

coordinates by first computing an initial harmonic parameterization (Floater 1997) and then applying a patch optimization technique (Sander et al. 2002).

The image on the left of Fig. 2 illustrates the resulting parameterization of our Rooster model. Each disk in the 2D texture map corresponds to a surface segment of the 3D model.

(b) *Texture Generation*: The texture map for the parameterized surface is computed in three steps:

1. **Identify regions of input images**: The objective of this step is to compute for each patch of the texture map (the disks in the second image from the left in Fig. 2) pixel colors, which accurately represent the surface colors of the 3D object at the corresponding points. This is achieved by projecting the corresponding surface patch, one triangle at a time, onto all input images where it is visible. We call the resulting section of the input image the back projection map and we call the resulting mapping between surface triangles and input image regions the back-projection mapping. The projection is only performed if the angle between a triangle's normal and the ray shooting from the triangle's centroid to the estimated camera position of the input image is larger than 90°.
2. **Texture map computation**: The image regions defined by the back projection map define the color information for the corresponding patch of the texture map. Using back projection mapping and the surface parameterization we can compute for each triangle of the surface mesh a mapping from the input image to the texture's parameter space. The algorithm is repeated for all patches of the reconstructed surface texture region and yields a set of overlapping textures covering the object.
3. **Minimize seams between overlapping textures**: Seams between overlapping textures are minimized by using a graph cut algorithm (Kwatra et al. 2003). We investigated different parameters settings for image fusion applications and found that Kwatra et al. cost function (gradient weighted color distance) in combination with the RGB color space and the L_2 norm works well for most applications (Clark et al. 2012).

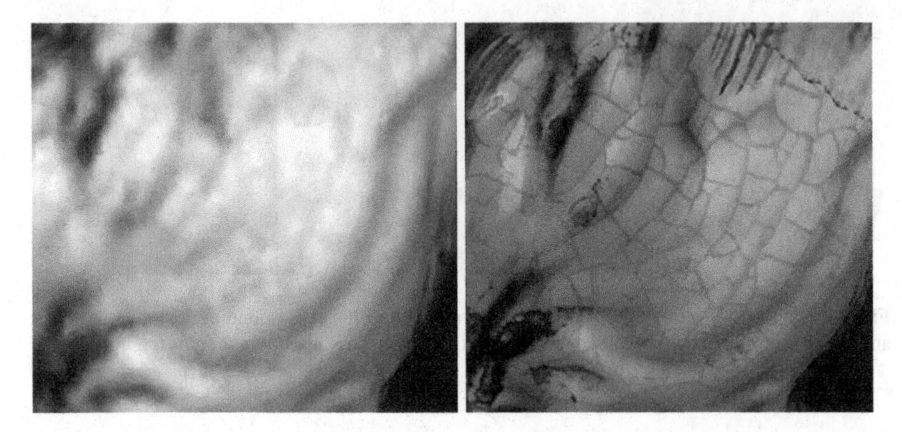

Fig. 3 Texture reconstruction by computing vertex colors and interpolating them (*left*) and the texture obtained using our approach (*right*). Note that both images show the neck section of the rooster in Fig. 2. The cracks in the image on the right reflect accurately the appearance of the object's material

The rightmost image in Fig. 2 shows the texture map obtained by back projection surface patches onto the input images and the resulting textured 3D model. Regions where no texture information was recovered are indicated in red. A typical reason is that users forget to make photos of the underside of the imaged object. In this case the Poisson surface reconstruction will still create a smooth surface interpolating the gaps in the point cloud, but no texture is available since that part of the surface is not shown on any input image. Figure 3 illustrates the level of detail obtainable with our texture reconstruction process.

4 Results

We tested our image-based modeling system using more than 40 data sets of both indoor and outdoor scenes, and of objects of different scale. Our system produces qualitatively good results for both uniformly colored and feature-poor objects, and for objects with concave regions and moderately complex geometries. The size of our test datasets varied from as few as 6 images to hundreds of images. All input images were acquired with simple consumer level cameras, including a Smartphone camera. The average computation time varies between 12 min to 10 h. Our system fails for objects which have viewpoint dependent surface appearance, e.g., refractive and reflective materials within complex environments. The following paragraphs present three examples of our results.

Fig. 4 Three out of 37 input images of the horse model data set (*left*) and the resulting reconstructed 3D model (*right*)

4.1 Horse Model

The dataset consists of 37 images of a wooden horse model. The images were acquired outdoors on a sunny day and have a resolution of 2592×1944 pixels. Three of the images are shown on the left of Fig. 4. The original object has a very smooth, reflective and shiny surface with few distinctive visual features. The resulting reconstructed model, shown on the right of Fig. 4, is of excellent quality and bears a high resemblance to the original object. The resulting model consists of 329,275 polygons and requires approximately 5 h and 12 min on an Intel Quad Core i7 with 6 GB RAM.

4.2 Miniature House Model

This dataset consists of 27 images of a replica of the famous house in Alfred Hitchcock's movie "Psycho". The images have a resolution of 2592×1944 pixels and were acquired with a consumer-level SONY DSCW18 camera under complex lighting condition (multiple spotlights and diffuse lights). The model's surface has a complex shape with many small features and holes.

The resulting reconstructed object (right hand side of Fig. 5) consists of 208,186 polygons and has an acceptable visual quality. The detailed fence-like structure on top of the roof and the tree leaves could not be accurately reconstructed since they were too blurry in the input images. Hence neither the shape-from correspondence approach, nor the shape-from-silhouette approach could create a sufficiently high number of points for capturing the 3D geometry. The computation time of this data set was 4 h 21 min on an Intel Quad Core i7 with 6 GB RAM.

Fig. 5 One of 27 input images of a miniature house model (*left*) and the resulting reconstructed 3D model (*right*)

Fig. 6 Two out of 21 input images of the elephant model data set (*left*) and the resulting 3D reconstruction (*right*)

4.3 Elephant Model

The elephant model consists of 21 images as illustrated on the left of Fig. 6. The images have a resolution of 2592 × 1944 pixels and were acquired with a consumer-level SONY DSCW180 camera in an indoor environment with relatively low light setting. The object has a complex surface geometry with many bumps and wrinkles, but few distinctive textural features. The resulting 3D reconstruction, shown on the right of Fig. 6, has 198,857 faces and is of very good quality. The texture and surface geometry of the object contain surprisingly accurate surface details. This example illustrates that our system performs well for objects with dark, rough surfaces and under relatively poor lighting conditions with large illumination variations and shadowing. The reconstruction process took almost 3 h to complete on an Intel Quad Core i7 with 6 GB RAM.

5 Conclusions and Future Work

We have described a novel image-based modelling system creating high quality 3D models fully automatically from a moderate number (20–40) of camera images. Input images are unconstrained and uncalibrated, which makes the system especially useful for low-cost and miniature mobile robots. In contrast to laser scanners our system also works for shiny and dark objects. The system still has some drawbacks which need to be addressed in future research. Missing regions in the texture map occur if the input images do not cover the entire object. We are currently working on texture inpainting techniques to fill these regions (Bertalmio et al. 2000; Perez et al. 2003).

Acknowledgements We would like to thank Prof. Eugene Zhang from the Oregon State University for providing us with code for his Feature-based Surface Parameterization technique (Zhang et al. 2005) and assisting with integrating it in our system.

References

Amenta N, Choi S, Kolluri RK (2001) The power crust. In: Proceeding of the 6th ACM symposium on solid modeling and applications. ACM Press, New York, pp 249–266

Baumgart BG (1974) Geometric modeling for computer vision. Doctoral Dissertation, Stanford University

Bernardini F, Mittleman J, Rushmeier H, Silva C, Taubin G (1999) The ball-pivoting algorithm for surface reconstruction. IEEE Trans Visual Comput Graphics 5(4):349–359

Bertalmio M, Sapiro G, Caselles V, Ballester C (2000) Image inpainting. In: Proceedings of the 27th annual conference on computer graphics and interactive techniques, pp 417–424

Cheng W, Ooi WT, Mondet S, Grigoras R, Morin G (2011) Modeling progressive mesh streaming: does data dependency matter. ACM Trans Multimedia Comput 7(2):1–24

Chien CH, Aggarwal JK (1984) A volume surface octree representation. In: Proceedings of the 7th international conference on pattern recognition, pp 817–820

Clark XB, Finlay JG, Wilson AJ, Milburn KLJ, Nguyen HM, Lutteroth C, Wunsche BC (2012) An investigation into graph cut parameter optimisation for image-fusion applications. In: Proceedings of the 27th conference on image and vision computing New Zealand, pp 480–485

Eck M, DeRose M, Duchamp T, Hoppe H, Lounsbery M, Stuetzle W (1995) Multiresolution analysis of arbitrary meshes. In: Proceedings of the 22nd annual conference on computer graphics and interactive techniques, pp 173–182

Edelsbrunner H (1995) Smooth surfaces for multi-scale shape representation. In: Proceedings of the 15th conference on foundations of software technology and theoretical computer science, pp 391–412

Floater MS (1997) Parametrization and smooth approximation of surface triangulations. Comput Aided Geom Des 14(3):231–250

Früh C, Zakhor A (2003) Constructing 3D city models by merging ground-based and airborne views. In: Proceedings of the IEEE inter conference on computer vision and pattern recognition 2:562–569

Grauman K, Shakhnarovich G, Darrell T (2003) A bayesian approach to image-based visual hull reconstruction. In: Proceedings of the IEEE inter conference on computer vision and pattern recognition 1:187–194

Henry P, Krainin M, Herbst E, Ren X, Fox D (2012) RGB-D mapping: using kinect-style depth cameras for dense 3D modeling of indoor environments. Int J Rob Res 31(5):647–663

Hernandez C, Vogiatzis G, Cipolla R (2008) Multi-view photometric stereo. Trans Pattern Recogn Mach Intell 30:548–554

Hilaga M, Shinagawa Y, Komura T, Kunii TL (2001) Topology matching for fully automatic similarity estimation of 3D shapes. In: Computer graphics proceedings, annual conference series, pp 203–212

Kazhdan M, Bolitho M, Hoppe H (2006) Poisson surface reconstruction. In Proceedings of the 4th Eurographics symposium on geometry processing, pp 61–70

Kwatra V, Schödl A, Essa I, Turk G, Bobick A (2003) Graphcut textures: image and video synthesis using graph cuts. ACM Trans Graph 22(3):277–286

Lorensen WE (1995) Marching through the visible man. In: Proceedings of the 6th conference on visualization, pp 368–373

Lowe DG (1999) Object recognition from local scale-invariant features. Proc Inter Conf Comput Vision 2:1150–1157

Lowe DG (2004) Distinctive image features from scale-invariant keypoints. Inter J Comp Vision 60:91–110

Martin W, Aggarwal JK (1983) Volumetric descriptions of objects from multiple views. Trans Pattern Anal Mach Intell 5(2):150–158

Matusik W, Buehler C, Raskar C, Gortler SJ, McMillan L (2000) Image-based visual hulls. In: Proceedings of the 27th annual conference on computer graphics and interactive techniques, pp 369–374

Melax S (1998) Simple, fast, and effective polygon reduction algorithm. Game Developer Magazine, pp 44–49

Newcombe RA, Izadi S, Hilliges O, Molyneaux D, Kim D, Davison AJ, Kohli P, Shotton J, Hodges S, Fitzgibbon A (2011) KinectFusion: real-time dense surface mapping and tracking. In: Proceedings of the 10th IEEE inter symposium on mixed and augmented reality, pp 127–136

Nguyen HM, Wunsche BC, Delmas P, Lutteroth C (2011) Realistic 3D scene reconstruction from unconstrained and uncalibrated images. In: Proceedings of the international conference on computer graphics theory and applications, 31:67–75

Oliver A, Kang S, Wünsche BC, MacDonald B (2012) Using the kinect as a navigation sensor for mobile robotics. In Proceedings of the 27th conference on image and vision computing, pp 509–514

Perez P, Gangnet M, Blake A (2003) Poisson image editing. ACM Trans Graph 22(3):313–318

Quan L, Tan P, Zeng G, Yuan L, Wang J, Kang SB (2006) Image-based plant modeling. ACM Trans Graph 25(3):599–604

Reeb G (1946) Sur les points singuliers dune forme de pfaff completement integrable ou diune fonction numerique (on the singular points of a completely integrable pfaff form or of a numerical function). Comptes Randus Acad Sciences Paris 222:847–849

Sander PV, Gortler SJ, Snyder J, Hoppe H (2002) Signal-specialized parameterization. In: Proceedings of the 13th Eurographics workshop on rendering, pp 87–100

Snavely N, Seitz S, Szeliski R (2006) Photo tourism: exploring photo collections in 3D. ACM Trans Graph 25(3):835–846

Szeliski R (2006) Image alignment and stitching. A tutorial in computer graphics and vision

Xiao J, Fang T, Tan P, Zhao P, Ofek E, Quan L (2008) Image-based façade modeling. ACM Trans Graph 27(5):26–34

Zhang E, Mischaikow K, Turk G (2005) Feature-based surface parameterization and texture mapping. ACM Trans Graph 24(1):1–27

Design of 3D Scene Scanner for Flat Surface Detection

A. Lipnickas, K. Rimkus and S. Sinkevičius

Abstract This chapter describes the design of a 3D space scene scanning system built from a 2D laser scanner merged with a CCD colour camera; it also presents an algorithm for flat area detection in a 3D point cloud. For that purpose, the RANdom SAmple Consensus (RANSAC) search engine has been adopted for flat area segmentation and planes detection. Due to a fact that human made planes are limited in size, we have proposed data filtering by comparing averaged point triangulation normals to the selected plane normal. The experiments have shown good results for an analysed environment segmentation with the applied angle variation measure up to ±25°. The applied variation threshold allowed to segment flat planes areas considering surface curvedness.

1 Introduction

Registration of three-dimensional data is essential for machine vision, object recognition, motion control and robot navigation applications. The main purpose of a visual sensing system is only to detect the presence of any obstacles; a more complex purpose of these machine vision systems is object detection and recognition. There are several techniques to perform sensing and measuring operations, but, depending on the technology used, they can be grouped into passive and active

A. Lipnickas · K. Rimkus (✉) · S. Sinkevičius
Department of Control Technology, Kaunas University of Technology, Kaunas, Lithuania
e-mail: kestas.rimkus@gmail.com

A. Lipnickas
e-mail: arunas.lipnickas@ktu.lt

S. Sinkevičius
e-mail: saulsink@gmail.com

Z. S. Hippe et al. (eds.), *Issues and Challenges in Artificial Intelligence*,
Studies in Computational Intelligence 559, DOI: 10.1007/978-3-319-06883-1_2,

sensor systems (Surgailis et al. 2011; Shacklock et al. 2006; Balsys et al. 2009). A passive sensor relies upon ambient radiation; an active sensor, however, illuminates the scene with radiation (often a laser beam or structured light) and determines how this emission is reflected.

Active laser scanning devices are known as LIDAR systems and available for 2D and 3D scene measuring. The operation principle of the LIDAR is based on measuring active signal time-of-flight (TOF) (Shacklock et al. 2006; Wulf and Wapner 2003; Surmann et al. 2001; Scaramuzza et al. 2007) later, the TOF is converted into a distance. To obtain 3D data, the beam is steered through an additional axis (tilt) to capture spherical coordinates {r, θ, φ: range, pan, tilt}. There are many examples on how to implement such systems (Himmelsbach et al. 2008): rotating prisms, polygonal mirrors, etc. As commercial 3D LIDAR systems are very expensive, many researchers convert commercial 2D laser devices into 3D ones by introducing an extra axis, either by deflecting the beam with an external mirror or by rotating a complete sensor housing (Surmann et al. 2001; Scaramuzza et al. 2007; Klimentjew et al. 2009). For some specific applications an adherent drawback of these systems is the absence of colour information on the points measured. Colour information would allow to detect all kinds of obstacles as well as occupancy of free areas.

When the speed of a 3D scene registration depends on the electromechanical properties of the scanning device, 3D data processing mostly depends on the scanned area and the density of the points measured. Usually, manipulation with 3D data is a very time consuming process, and it requires a lot of computing resources, because three dimensional scenes or objects can consist of thousands to millions of measured points.

Over the last decade, despite the development of new scanning devices, there has been growing interest in the development of new methods which would allow to reduce the number of constructive primitives without a visible loss of the geometrical form or shapes of the objects scanned (Joochim and Roth 2008; Aguiar et al. 2010). Curvature is one of the most important characteristics of the 3D surface and it is an indicator of the sharpness of a 3D object. Non ridge surfaces or sectors of flat surfaces are mostly non informative for the recognition, however, they are scanned with the same measuring point density. These areas are redundant and can be eliminated with the goal of to reducing the amount of data, economizing on the use of computing resources and preserving geometrical accuracy at the same time.

In this chapter, we propose a 3D range scanner composed of a 2D laser scanner with an extra gear supplying the third degree of freedom as well as a charge-coupled device (CCD) camera for colour information perception. The advantages of the technique proposed in this chapter are the following: usage of a calibration technique to couple the CCD camera with the 3D laser scanner as well as application of the RANSAC method for non-informative plane detection and descriptive 3D primitives reduction.

This chapter is organized as follows. In the next section the state of the art is given, followed by a detailed description of the system setup. Section 4 provides the results of the plane detection. Section 5 concludes the chapter and gives an outlook for the future research.

2 State of the Art

The idea of reducing the number of the constructive primitives of a 3D object has gained an increasing interest in today's computer vision, pattern recognition and image processing research field. In Hinker and Hansen (1993) the authors have proposed a method that combines coplanar and nearly coplanar polygons into a larger polygon which can be re-triangulated into fewer simpler polygons. The geometric optimization described in Hinker and Hansen (1993) performs best on geometries made up of larger objects; for example, isosurfaces generated from three dimensional hydrodynamics simulations and/or 3D range scanners. Kalvin and Taylor (1996) have presented a simplification method called "superface". The superface, like the method mentioned above, simplifies polygons by merging coplanar faces and then triangulating the perimeter of these merged faces. The merging procedure is controlled by considering an infinite number of possible approximating plane solutions. The merging of a new face into current superfaces is stopped when a solution set of approximating planes disappears. Garland et al. (2001) have presented a method which can be applied not only to surface simplification tasks, but also to the detection of collision and intersection. This method is based on iterative clustering, i.e., pairwise cluster merging. Such a merging technique produces a hierarchical structure of clusters.

Another technique for the representation of highly detailed objects is presented by Lee et al. (2000). They have defined the domain surface using subdivision surfaces which are applied as descriptors for the representation of smooth surfaces. The displaced subdivision surface consists of a control mesh and a scalar field that displaces the associated subdivision surface locally along its normal vectors. The authors of Lee et al. (2000) have demonstrated that this method offers significant storage savings.

3D object detection algorithms are often divided into two groups: *model-based* and *model-free*. Model-based approaches try to solve detection and classification simultaneously by fitting models to the data. The novelty of the technique presented in this work is based on adoption of the RANSAC technique for detecting big flat areas based on a *plane-model* with redundant points by incorporating information of constructive primitives (colour and/or tangential normal). The described technique has been successfully applied to a scene consisting of various 3D objects to eliminate flat areas such as walls, flooring, boards etc.

3 Experimental Setup

In our work, we use a 2D scanning laser range finder UBG-04LX-F01 for area scanning with laser class 1 safety of wavelength 785 nm (Fig. 1); a servomotor was used for supplying rotational movement and a video camera for supplying colour information of a visible scene.

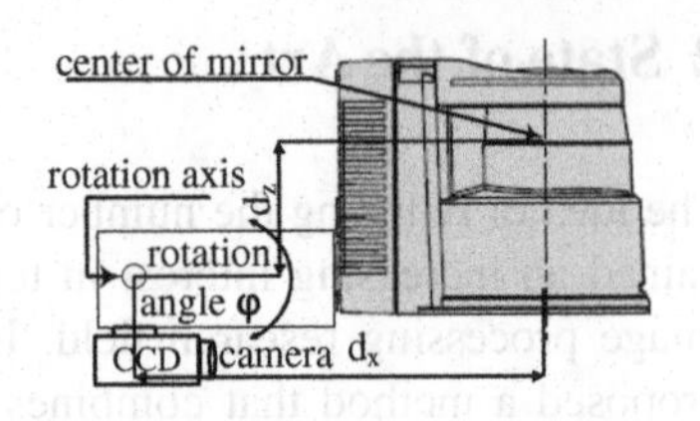

Fig. 1 Schematic of the 3D sensor

The scan area of the 2D laser scanner is a 240° semicircle with the maximum radius of 4 m. The pitch angle is 0.36° and the sensor outputs the distance measured at every point (682 steps). Accuracy ranges are: in range of 0.06–1 m: ±10 mm, and in range of 1–4 m: 1 % of distance (Hinker and Hansen 1993). The servomotor Hitec-HS-422HD with step angle 0.02° and torque 3.0 kg cm is used to supply an additional coordinate φ (see Fig. 1).

The designed system integrates a digital camera (type FFMV-03MTC-60), mounted just below the central axis of a laser range finder. Because of two independent systems used for colour 3D scene scanning, they had to be calibrated to match the same measured and sensed points. The calibration procedure is presented in the next section.

The accuracy of 3D data colour mapping mostly depends on the accuracy of camera calibration. To estimate the camera's parameters we have applied Jean-Yves Bouguet's well-known and widely-used Camera Calibration Toolbox (Bouguet 2006) in our work.

The most widely used model of the camera is a pinhole model (Joochim and Roth 2008; Serafinavičius 2005). The equation of the camera model is (1):

$$\begin{bmatrix} u \\ v \\ 0 \end{bmatrix} = \frac{f}{z} \cdot \begin{bmatrix} k_u & 0 & \frac{u_0}{f} \\ 0 & k_v & \frac{v_0}{f} \\ 0 & 0 & \frac{1}{f} \end{bmatrix} \cdot \begin{bmatrix} x \\ y \\ z \end{bmatrix}, \tag{1}$$

where u, v are the image plane coordinates, x, y, z are the world coordinates, k_u, k_v are scale factors along the axes of pixel coordinates, u_0, v_0 are the pixel coordinates of the principal point (orthogonal projection of the optical centre on the image plane), f is the focal length.

3D laser range finder data are expressed in spherical coordinates. The sensor model can be written as (Wulf and Wapner 2003):

$$\begin{bmatrix} x \\ y \\ z \end{bmatrix} = \begin{bmatrix} c_i c_j & -c_i c_j s_j & s_i & c_i d_x + s_i d_z \\ s_j c_i & c_i & 0 & 0 \\ s_i & s_i c_i s_j & c_i & s_i d_x + c_i d_z \end{bmatrix} \begin{bmatrix} \rho_{ij} \\ 0 \\ 0 \\ 1 \end{bmatrix}, \tag{2}$$

where, $c_i = \cos(\varphi_i)$, $c_j = \cos(\theta_j)$, $s_i = \sin(\varphi_i)$, $s_j = (\theta_j)$, ρ_{ij} is the j-th measured distance with corresponding orientation Θ_j in the i-th scan plane, which

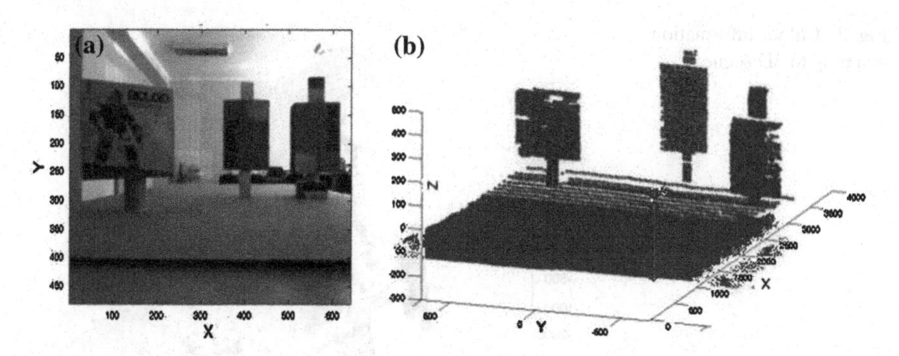

Fig. 2 Undistorted image (**a**) and 3D scanned scene (**b**)

makes the angle φ_i with the horizontal plane (Fig. 1). The offset of the external rotation axis from the centre of the mirror in the laser frame has components $d_x = 90$ [mm] and $d_z = 20$ [mm]. The $[x, y, z]^T$ are the coordinates of each measured point relative to the global frame (with its origin at the centre of the rotation axis, the x-axis pointing forward and the z-axis toward the top).

In the case of mapping $[x, y, z]^T$ point with its colour, it is possible to add depth value (z) to the visual image (u, v) and, conversely, it is possible to add point colour information to the measured data points (x, y, z). The only difficulty here is to find the corresponding formula which would map respective points in both measuring systems.

In order to map camera points with a laser scanner system, the tangential and radial distortion of the camera had to be corrected. The calibration and parameters of the internal camera model were determined according to the instruction given in Bouguet (2006). Camera calibration results (with uncertainties) are as follows:

- Focal Length: $[f_{cx}, f_{cy}] = [\,490.23\ 488.58\,] \pm [\,0.72\ 0.71\,]$;
- Principal point: $[a_{u0}, a_{v0}] = [\,261.76\ 233.23\,] \pm [\,0.49\ 0.51\,]$;
- Distortion:
  $d_c = \left[-0.36235\ 0.144\ 0.001\ -0.001\ 0.000\right] \pm \left[0.001\ 0.002\ 0.001\ 0.001\ 0.000\right]$;
- Pixel error: $\mathrm{err} = [\,0.164\ 0.169\,]$.

The estimated distortion parameters allow correcting the distortion in the original images. Figure 2 displays a distortion-free image and a 3D scanned scene.

The same undistorted image (Fig. 2a) is used for mapping 2D visual information to a 3D scanned area shown in Fig. 2b. For the mapping, 24 points on box corners in the scene were selected in 2D images and 3D point clouds.

The data mapping model is the classical pinhole camera model described in (1). The parameters of this model were obtained by applying the least squares (lsq) fitting technique. Following a lengthy application of the trial and error method to determine the model and its parameters, the following optimal solution was obtained: the mean-average–error MAE $= 1.9$ and mean-square-error

Fig. 3 Colour information mapping to 3D scene

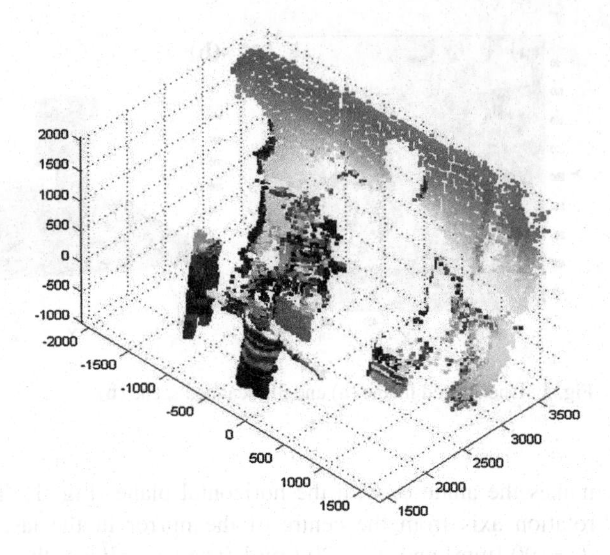

MSE = 6.62. The mapping functions for a 2D image plane (u, v) are with the corresponding coefficient vectors c_u and c_v:

$$c_u = \left[\, 111.2 \; 479.1 \; 254.7 \; 39.0 \,\right],$$

$$c_v = \left[\, 175.8 \; 550.4 \; 235.3 \; 19.8 \,\right],$$

$$u_i = c_u(3) + \frac{c_u(2) \cdot (x_i + c_u(1))}{z_i - c_u(4)}, \tag{3}$$

$$v_i = c_v(3) + \frac{c_v(2) \cdot (y_i + c_v(1))}{z_i - c_v(4)} \tag{4}$$

where, the first coefficient of c_u and c_v describes a physical mounting mismatch in millimetres between the centres of 3D and 2D systems, the second parameter describes a new focal length similar to f_{cx}, f_{cy}, the third parameter describes a new principal point a_{u0}, a_{v0}, and finally the fourth parameter describes the correction of the depth measurement. The mismatch between the camera calibration results and mapping model parameters are due to the fact that the undistorted images are not the same size as used for calibration. With the models obtained (in Eqs. 3, 4) it becomes possible to map each measured 3D point (x_i, y_i, z_i) to a corresponding point in the image plane (u_i,v_i) and vice versa (see Fig. 3). As it is seen from Figs. 2 and 3, man-made scenes usually consist of big and flat areas. In order to reduce the amount of computation for data analysis, flat areas can be removed as non-informative parts (Aguiar et al. 2010). For that purpose, the RANSAC search technique was applied for plane detecting in a 3D data point cloud.

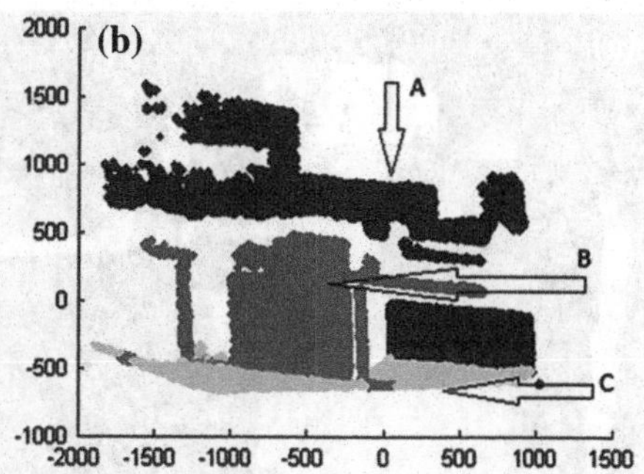

Fig. 4 Gray scene view analyzed by the RANSAC method (**a**) and the detected planes {A, B, C} (**b**)

4 Planes Estimation Using the RANSAC

The observed plane in a 3D point cloud is described by plane equation:

$$C(1) * X + C(2) * Y + C(3) * Z + C(4) = 0 \tag{5}$$

where, C is 4×1 array of plane coefficients.

For the purpose of detecting flat planes we have used a scanned 3D scene shown in Fig. 4a. The result was three fitted planes shown in Fig. 4b. As it is seen from Fig. 4b, plane A corresponds to the wall, C—to the floor and plane B corresponds to the front of the cabinet. Physically, plane B is not a single plane since its points lie on two desks and the cabinets. The challenge here is how to avoid the assignment of desk points to the same plane as the front of the cabinets. The answer is to measure the curvedness of the point neighbourhood or compare the face's normals of triangulated elements to the normal of the defined plane.

A strong causal relationship between the measured curvedness of the scanned 3D scene and the computed plane was not determined due to considerable noise which accompanied the measuring procedure. But the comparison of point triangulation normals to the normal of the plane allows the separation of desk points from the cabinet points (Fig. 5). The angle is measured as a dot product of two vectors, i.e. plane normal and point triangulation normal. Point triangulation normal is calculated as the average of all triangulated neighbourhood elements connected to that point.

Depending on the selection of the filtering threshold, the number of selected points will vary as well e.g., by making a small threshold gate for possible variation of triangulated normals to the measured plane, only a small portion of data points will be assigned; and vice versa, for too higher threshold, no points will be filtered out. Therefore, the investigation of data filtering based on the angle variation has been carried out. Figure 5 shows the plane B of the scene in Fig. 4a before (a) and after (b) data filtering with acceptable variation up to $\pm 25°$.

Figure 6 shows the dependency between angle variation (abscissa) in degrees and the remaining points after filtering out the data in percentage (ordinate). The

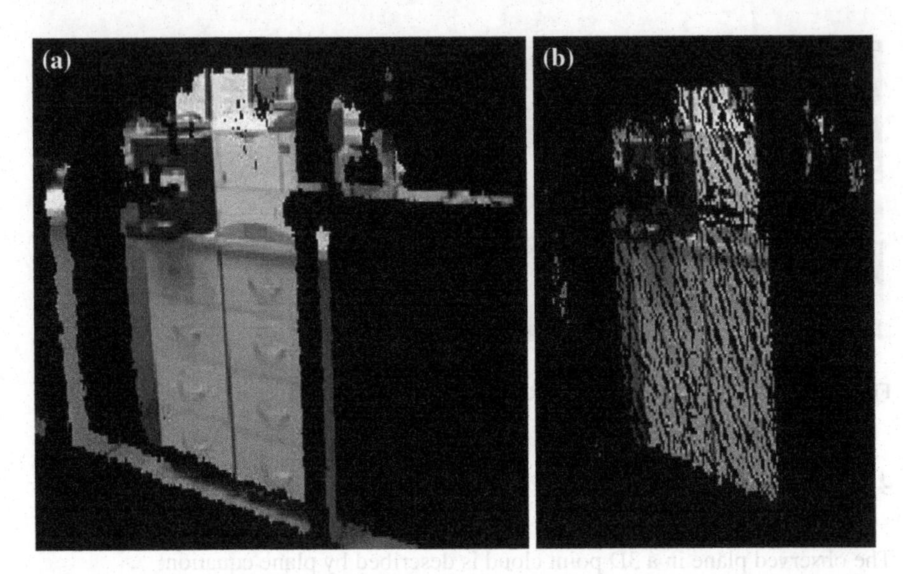

Fig. 5 Detected plane B by RANSAC method (**a**) and in (**b**) the plane B with filtered out normal's bigger than ±25°

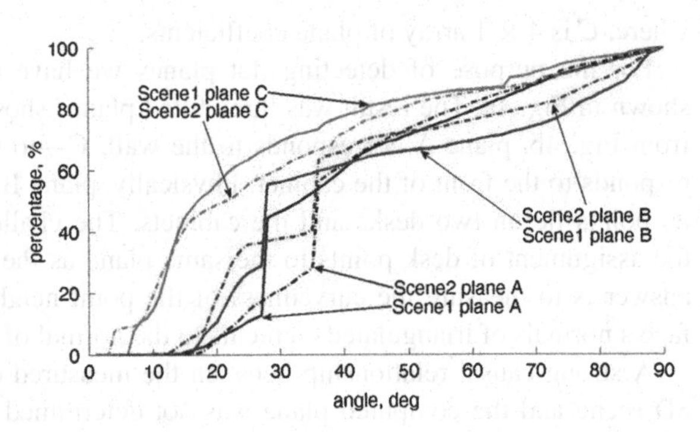

Fig. 6 The dependency between angle variation (*abscissa*) in degrees and remaining points after filtering out the data in percentage (*ordinate*)

curves in the graphs correspond to the planes from two scenes. As it is seen from Fig. 6, curve behaviour largely depends on the nature of the scene and objects presented in the scene.

From our experience we can say that the most appropriate angle variation is acceptable only up to ±30°.

5 Discussion and Conclusions

In this chapter, we have presented a 3D space scene scanning system built of the 2D laser scanner and the CCD camera. A mapping function for colour information mapping is derived. The designed system allows straightforward building of colour depth maps as well as assigning colour information to 3D points measured.

The RANSAC method applied for the planes detection in 3D point clouds always gives some planes found in the scene. In reality, these scenes do not continue into infinity; therefore, points from different objects are automatically assigned to the plane. In this chapter we are proposing data filtering method based on the comparison of averaged point triangulation normals to the selected one. The results have shown that due to the scanning noise of the 3D scanning system, the acceptance threshold should be set to ±30°. The presented approach can be applied for solving extraction problems of the constructive primitive, i.e. the problem of the topological simplification of the scanned scenes. It can be used as part of complex solutions for surface re-meshing tasks as well as for reducing constructive primitives. A simplified 3D object consisting of the reduced number of constructive primitives requires less computation and the storage.

Future work will consist of augmenting the discrimination properties of our method with a classification framework for 3D object identification.

Acknowledgment This research is funded by the European Social Fund under the project "Microsensors, microactuators and controllers for mechatronic systems (Go-Smart)" (Agreement No VP1-3.1-ŠMM-08-K-01-015).

References

Aguiar CSR, Druon S, Crosnier A (2010) 3D datasets segmentation based on local attribute variation. In: IEEE/RSJ international conference on intelligent robots and systems, pp 3205–3210

Balsys K, Valinevičius A, Eidukas D (2009) Urban traffic control using IR video detection technology. Electron Electr Eng—Kaunas: Technologija 8(96):43–46

Bouguet J (2006) Camera calibration toolbox for matlab. http://www.vision.caltech.edu/bouguetj/calib_doc. Accessed 5 Dec 2013

Garland M, Willmott A, Heckbert PS (2001) Hierarchical face clustering on polygonal surfaces. In: Proceedings of ACM symposium on interactive 3D graphics, pp 49–58

Himmelsbach M, Muller A, Luttel T, Wunsche HJ (2008) LIDAR-based 3D object perception. In: Proceedings of 1st international workshop on cognition for technical systems

Hinker P, Hansen C (1993) Geometric optimization. Visualization 93:189–195

Joochim C, Roth H (2008) Development of a 3d mapping using 2d/3d sensors for mobile robot locomotion. In: Proceedings of IEEE international conference of techologies for practical robot applications, pp 100–105

Kalvin A, Taylor R (1996) Superfaces: polygonal mesh simplification with bounded error. IEEE Comput Graphics Appl 16:64–77

Klimentjew D, Arli M, Zhang J (2009) 3D scene reconstruction based on a moving 2D laser range finder for service-robots. In: Proceedings of IEEE international conference on robotics and biomimetics, pp 1129–1134

Lee A, Moreton H, Hoppe H (2000) Displaced subdivision surfaces. In: Proceedings of 27th annual conference on computer graphics and interactive techniques, pp 85–94

Scaramuzza D, Harati A, Siegwart R (2007) Extrinsic self calibration of a camera and a 3d laser range finder from natural scenes. In: Proceedings of IEEE international conference on intelligent robots and systems, pp 4164–4169

Serafinavičius P (2005) Investigation of technical equipment in computer stereo vision: camera calibration techniques. Electron Electr Eng—Kaunas: Technologija 3(59):24–27

Shacklock A, Xu J, Wang H (2006) Visual guidance for autonomous vehicles: capability and challenges. In: Shuzhi SG, Frank LL. Autonomous mobile robots: sensing, control, decision making and applications, CRC Press, pp 8–12
Surgailis T, Valinevičius A, Eidukas D (2011) Stereo vision based traffic analysis system. Electron Electr Eng—Kaunas: Technologija 1(107):15–18
Surmann H, Lingemann K, Nuchter A, Hertzberg J (2001) A 3D laser range finder for autonomous mobile robots. In: Proceedings of 32nd international symposium on robotics, pp 153–158
Wulf O, Wapner B (2003) Fast 3d scanning methods for laser measurement systems. In: International conference on control systems and computer science, vol 1, pp 312–317

Validation of Point Cloud Data for 3D Plane Detection

A. Lipnickas, K. Rimkus and S. Sinkevičius

Abstract There are number of plane detection techniques for a given 3D point cloud utilized in different applications. All of the methods measure planes quality by computing sum of square error for a fitted plane model but no one of techniques may count the number of planes in the point cloud. In this chapter we present new strategy for validating number of found planes in the 3D:point cloud by applied cluster validity indices. For a planes finding in point cloud we have engaged the RANdom SAmple Consensus (RANSAC) method to synthetic and real scanned data. The experimental results have shown that the cluster validity indices may help in tuning RANSAC parameters as well as in determination the number of planes in 3D data.

1 Introduction

In recent years, various devices have been developed as an attempt to access the 3D information of the physical world, such as time-of-light (TOF) camera, a stereo camera, a laser scanner, and a structured light camera. With each new generation of these devices, they are becoming faster, more accurate and higher resolution. This means more data points in each frame, longer processing time (to object segmentation and identification). As example in 2004 presented (Gokturk et al. 2004)

A. Lipnickas
The Mechatronics Centre for Studies, Information and Research, Kaunas, Lithuania
e-mail: arunas.lipnickas@ktu.lt

K. Rimkus (✉) · S. Sinkevičius
Department of Control Technologies, Kaunas University of Technology, Kaunas, Lithuania
e-mail: kestas.rimkus@gmail.com

S. Sinkevičius
e-mail: saulsink@gmail.com

Z. S. Hippe et al. (eds.), *Issues and Challenges in Artificial Intelligence*,
Studies in Computational Intelligence 559, DOI: 10.1007/978-3-319-06883-1_3,

Time-Of-Flight (TOF) camera obtains 64 × 64 frame size and 2010 introduced to the market the Kinect sensor by Microsoft (2013) or Xtion Pro Live by Asus (2013) has a 480 × 640 frame size, that's mean 75 times more data points in each frame.

In object recognition task from a point cloud data the critical step is in extracting meaningful information by removing excessive information. A curvature is one of the most important characteristic of a 3D surface and it is an indicator of the sharpness of the 3D object. Non ridge surfaces or sectors of flat surface are mostly non informative for the recognition, but are scanned with the same measuring point density. Therefore these areas are redundant and can be eliminated for the data point reduction purpose and computational burden reduction at the same time preserving geometrical accuracy.

Flat surface in 3D point cloud can be found by using a mathematical plane model. But there is always the question—"How many planes are in the unknown data set?" and "How good these planes are detected?" In this chapter we present a new strategy to use planes as clusters and cluster validity indexes to tune and verify the planes detection procedure.

2 State of the Art

Literature review most often presents the plane detection methods in 3D point cloud based on: least square (LS), principal component analysis (PCA), region growing (plane growing), RANSAC, 3D Hough transform. Planes detection is done for variety of purposes for instance (Pathak et al. 2010a, b; Junhao et al. 2011) 3D planes are used as markers for a Simultaneous Localisation and Map building (SLAM) problem solving. Bare Earth surface and roof plane detection from airborne laser scanning data are analyzed in Tarsha-Kurdi et al. (2008), Huang and Brenner (2011), Vosselman (2009). Unfortunately, the final evaluation of the planes detection method quality and number of detected planes is done by the human experts.

As found in Schnabel et al. (2007) RANSAC algorithm performs precise and fast plane extraction with fine-tuned parameters even in large data set. We have adopted this method for planes detection in 3D point cloud. The numbers of parameters usually are adopted empirically. For the precise parameters tuning we have applied the cluster validity indices. Planes finding in 3D point cloud can be seen as data clustering with the same demand on the clustered data, i.e. to be compact and well separated. Therefore we see the cluster validity indices as the special tool to measure the number of planes in 3D point cloud as well as the validation technique for planes detection engine.

The cluster validity indices have been invented to validate clustering techniques and compactness of clustered data. There are variety of proposed cluster validity indices presented in Ansari et al. (2011), Halkidi et al. (2001), Milligan (1996) but only some of them poses clear output range suitable for analysis. In this work we have engaged cluster validity indices such as: Silhouette, Dunn's and Davies–Bouldin index; they are explained in next section.

3 Planes Detection in 3D Point Cloud and Validation Indexes

The RANSAC algorithm proposed by Fischler and Bolles (1981) is a general parameter estimation approach designed to cope with a large proportion of outliers in the input data. Unlike many of the common robust estimation techniques such as M-estimators and least-median squares that have been adopted by the computer vision community from the statistics literature, the RANSAC has been developed by the computer vision community researchers. The RANSAC is a resampling technique that generates candidate solutions by using the minimum number of observations (data points) required to estimate the underlying model parameters. As pointed out by Fischler and Bolles (1981), unlike conventional sampling techniques that use as much of the data as possible to obtain an initial solution and then proceed to prune outliers, RANSAC uses the smallest set as possible and proceeds to enlarge this set with consistent data points. Steps of RANSAC Algorithm:

1. Select randomly the minimum number of points required to determine the model parameters,
2. Solve for the parameters of the plane model,
3. Determine how many points from the set of all points fit with a predefined tolerance *t*,
4. If the fraction of the number of inliers over the total number points in the set exceeds a predefined threshold *t*, re-estimate the model parameters using all the identified inliers and terminate,
5. Otherwise, repeat steps 1 through 4 (maximum of *l* times). The number of iterations, *l*, is chosen high enough to ensure that the probability (usually set to 0.99) that at least one of the sets of random samples does not include *o* outlier.

The observed plane in 3D point cloud is described by the following equation:

$$\mathbf{C}(1) * \mathbf{X} + \mathbf{C}(2) * \mathbf{Y} + \mathbf{C}(3) * \mathbf{Z} + \mathbf{C}(4) = 0, \tag{1}$$

where **C**–4 × 1 array of plane coefficients.

3.1 Dunn's Validity Index

Dunn's Validity Index (Dunn 1974) attempts to identify those cluster sets (in this chapter points from planes) that are compact and well separated. For any number (k) of clusters, where c_i—represent the i-cluster of such partition, the Dunn's validation index (D) can be calculated with the following formula:

$$D = \min_{1 \le i \le k} \left(\min_{i+1 \le j \le k} \left(\frac{dist(c_i, c_j)}{\max\limits_{1 \le l \le k} (diam(c_l))} \right) \right) \tag{2}$$

where $dist(c_i, c_j)$ is the distance between clusters, $dist(c_i, c_j) = \min_{x_i \in c_i, x_j \in c_j} (d(x_i, x_j))$, $d(x_i, x_j)$ is the distance between data points $x_i \in c_i$ and $x_j \in c_j$, $diam(c_l)$ is a diameter of the cluster c_l, where $diam(c_l) = \max_{x_{l_1}, x_{l_2} \in c_l} (d(x_{l_1}, x_{l_2}))$ (Dunn 1974).

An optimal value of the k (number of clusters) is one that maximizes the Dunn's index.

3.2 Davies–Bouldin Validity Index

This index attempts to minimize the average distance between each cluster (planes) and the one most similar to it (Davies and Bouldin 1979). It is defined as:

$$DB = \frac{1}{k} \sum_{i=1}^{k} \max_{1 \le j \le k, j \ne i} \left(\frac{diam(c_i) + diam(c_j)}{dist(c_i, c_j)} \right) \tag{3}$$

An optimal value of the k is the one that minimizes this index.

3.3 Silhouette Validity Index

The silhouette value for each point is a measure of how similar that point is to points in its own cluster (planes) compared to points in other clusters. This technique computes the silhouette width for each data point, average silhouette width for each cluster and overall average silhouette width for the total data set (Rousseew 1987). Let be the data set are clustering into k clusters. To compute the silhouettes width of ith data point, following formula is used:

$$s_i^j = \frac{b_i^j - a_i^j}{\max\left(a_i^j, b_i^j\right)}, \tag{4}$$

where: a_i^j is average distance from the ith data point to all other points in the same (j) cluster; b_i^j is minimum of average distances of ith data point to all data points in others clusters (Kaufman and Rousseeuw 1990). Equation (4) results in $-1 \le s_i^j \le 1$. A value of s_i^j close to 1 indicates that the data point is assigned to a very appropriate cluster. If s_i^j is close to zero, it means that that data pint could be assign to another closest cluster as well because it is equidistant from both the clusters. If s_i^j is close to -1, it means that data point is misclassified and lies somewhere in between the clusters.

From the expression (4) we can now define the silhouette of the cluster c_j with size of m_j:

$$S_j = \frac{1}{m} \sum_{i=1}^{m_j} s_i^j. \tag{5}$$

The overall average silhouette width for the entire data set is the average S_j for all data points in the whole dataset. The largest overall average silhouette indicates the best clustering. Therefore, the number of cluster with maximum overall average silhouette width is taken as the optimal number of the clusters.

Finally, the global Silhouette index of the clustering is given by:

$$GS = \frac{1}{k}\sum_{j=1}^{k} S_j. \tag{6}$$

As the Authors of Ansari et al. (2011) conclude that there are number of cluster validity indices, but mostly are used indices having well defined output boundaries. The plane detection in 3D point cloud is very similar to data clustering problem with one exception, i.e. not assigned data points to N found planes remains unlabeled and neglected in index computation. In 3D point cloud the unassigned data can be seen as unstructured points or point without flat areas, therefore we propose compute the *GS* index only on data assigned to N planes and ignoring the unlabeled data. We call this index as *GSplanes*.

4 Experiments

For planes finding in 3D point cloud we have applied the RANSAC method to synthetic data (generated 3 perpendicular cross planes, i.e. one corner of cube), scanned 3D cube with six flat sides and 3D "Stanford bunny" data. For all data we calculate Dunn's (D), Davies–Bouldin (DB), Global Silhouette (GS) and GSplanes validity indices. In most cases the unassigned data to any known plane is labelled as one cluster; for calculation of GSplanes validity index the unassigned data points are ignored. The presented experiments were cared out for two main purposes: the first one—to find optimal predefined tolerance t used in RANSAC algorithm and for the second—to validate exact number of planes in the 3D point cloud data. The applied RANSAC algorithm searches plains iteratively, one by one. The new plain is always formed on remaining data after removing the data assigned to the known plane. Plain finding procedure is repeated while remaining data satisfy plane assignment criterion based on point tolerance t. Point tolerance t mostly depends on measurement accuracy or noise in measurement. The tolerance t might be assigned empirically or evaluated experimentally by the help of validation indices.

4.1 Synthetic 3D data

As a synthetic 3D data we generated three perpendicular cross ideal planes (plane model with three known noise variations e = [±0, ±2.5, ±7.5] of measuring

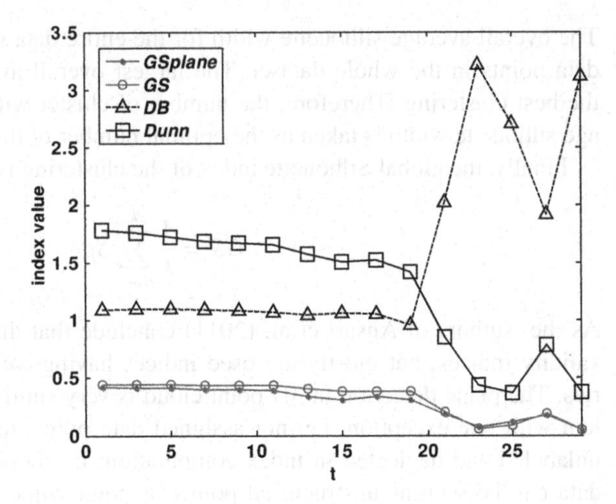

Fig. 1 Validity indexes values with various t values in case of $e = 0$ for synthetic data

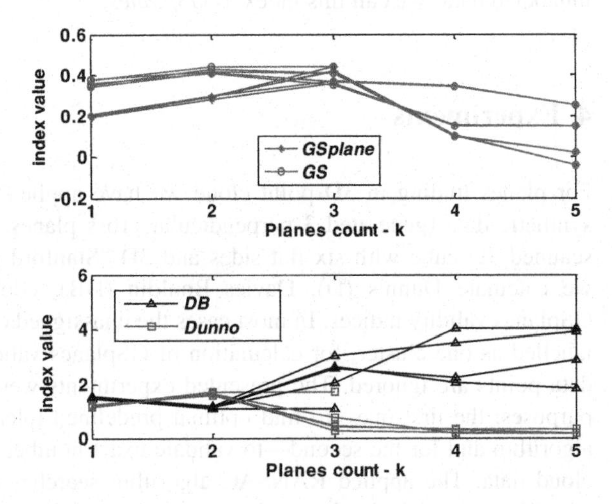

Fig. 2 Validity indexes values with various plane count with e = 0 on synthetic data from five trails

units) as a test data to find optimal predefined tolerance t for RANSAC algorithm by cluster validity indices. Figure 1 presents the calculated cluster validity indices on synthetic data with e = ±0 additional noise. From the Fig. 1 it is obvious that the max Dunn, Gs, GSplanes and min DB points out the best validity indexes values at t close to zero.

With determined t it is possible to run experiment for finding number of planes in 3D point cloud data. The RANSAC was run to find up to five planes on the synthetic data. Figure 2 shows the values calculated by *GS*, *GSplanes*, *DB* and *D* indexes on various number of planes (clusters) from five trails.

The *GS*, *DB* and *D* points out that the best number of fitted planes is 2 (having in mind that 2 planes are identified and the third one is the plane consisting

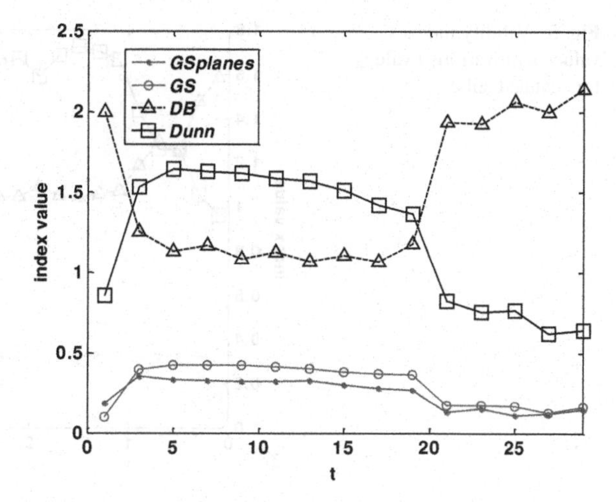

Fig. 3 Validity indexes values with various t in case of $e = 2.5$ for artificial data

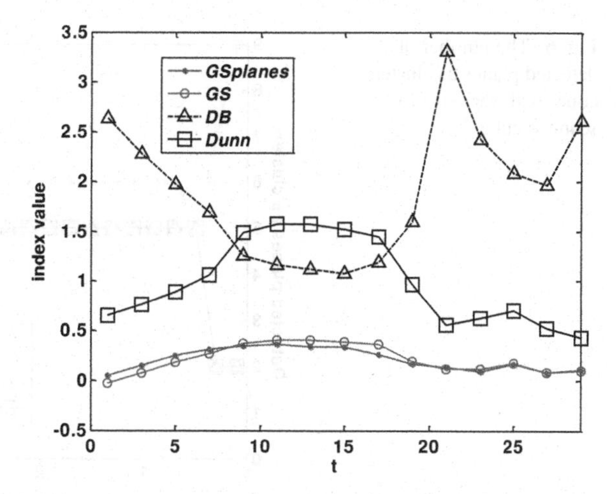

Fig. 4 Validity indexes values with various t values in case of $e = 7.5$ for artificial data

of remaining points). Only *GSplanes* points out exactly at 3 planes meaning that selected $t \approx 0$ from Fig. 1 is a good choice for this data set. We have repeated this experiment on synthetic data by adding noise $e = \pm 2.5$ and ± 7.5. Figures 3 and 4 show index values for various t on data with noise $e = \pm 5$ and ± 7.5 respectively. As it is seen from Fig. 3 if the noise is in range $e = \pm 2.5$ of measured units, then the best point tolerance parameter t should be selected from the range [2.5–15] and similarly if $e = \pm 7.5$, the t should be from the range [7.5–15].

Practically it means that the useful point tolerance parameter for RANSAC method should be set up to twice of data noise value. If data noise level is unknown, it might be evaluated by the help of cluster validity indices.

For the selected value of t the RANSAC finds the number of planes similarly to the case shown in Fig. 2.

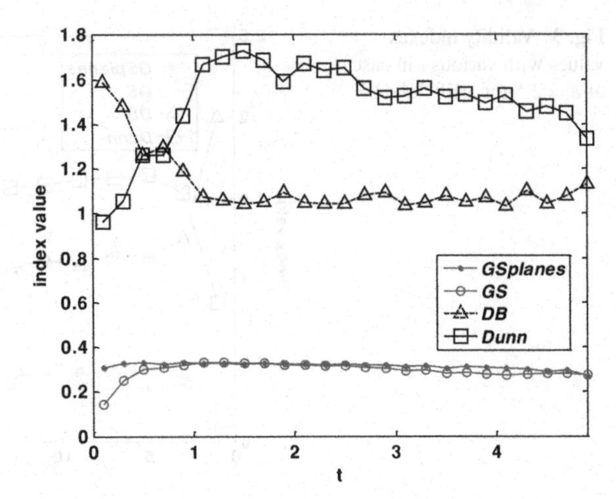

Fig. 5 Validity indexes values with various t values for scanned cube

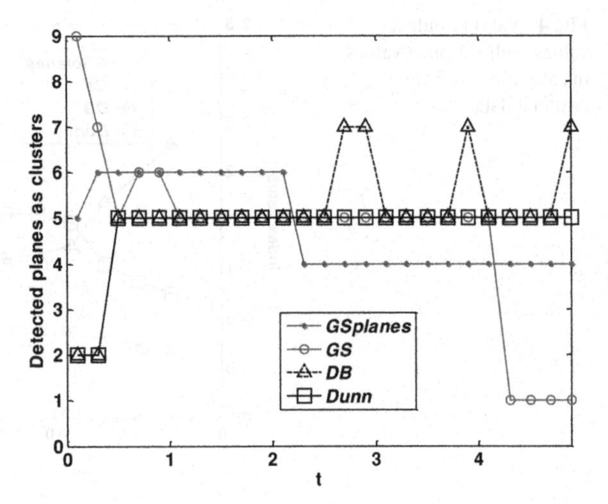

Fig. 6 The number of detected planes as clusters count with various t for scanned cube

4.2 Scanned Cube

As next test object we have took real scanned cube 3D data with unknown scan precision. As it is seen from Fig. 5 the best point tolerance is in range $t \approx 1 \div 2$ of measuring units. The number of most often found planes by the help of validity indices is five, only GSplanes points out that the stable number of planes is six. GSplanes index shows the best measuring accuracy for six planes object with rounded edges (due to measuring errors).

The Fig. 6 proves that the stable identification of number of planes is identified in data if parameter t is in the range of $[1 \div 2]$. As well as Fig. 6 demonstrate situation when t is selected too small or too big the plains are identified incorrectly.

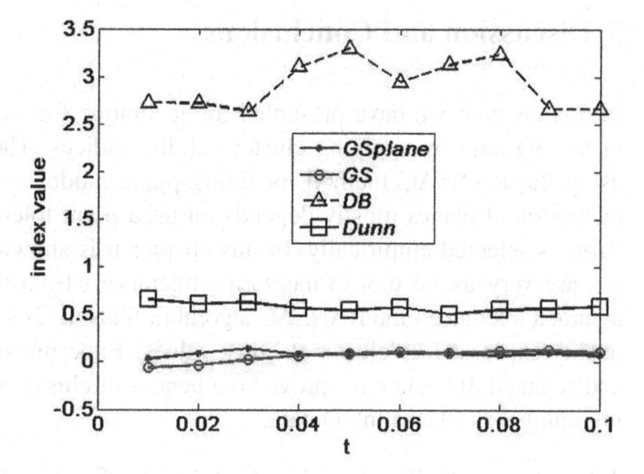

Fig. 7 Validity indexes values with various *t* values for "Stanford bunny" model

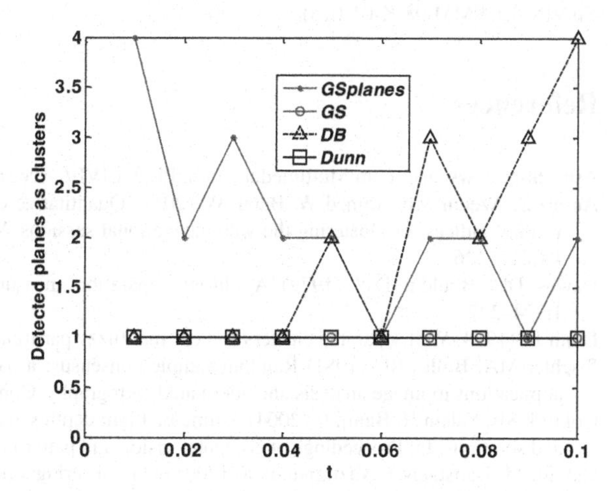

Fig. 8 Detected planes as clusters count with various *t* for "Stanford bunny" model

4.3 Bunny Model

For the last test run we have used "Stanford bunny" (Turk and Levoy 1994) without any expressed flatness on the scanned surface. As in previous runs we experimented with t variation from 0.01 till 0.1. As it is seen from Fig. 7 validation indices do not distinguishes any flatness in data.

As well as *GS* and *GSplanes* index values are very close to zero meaning that found planes are assigned in wrong way. Therefore from this experiment is impossible to select best *t* for planes analysis. In addition, Fig. 8 shows that the validity indexes disagree on possible number of planes in "Stanford bunny" data. Manual analysis of data has shown that there is single flatness i.e. the bottoms of bunny claws. The *GS* and *Dunn* indices, presented in Fig. 8, show that the number of stable planes is one.

5 Discussion and Conclusions

In this chapter, we have presented the technique for validating number of planes in the 3D data by applying cluster validity indices. The planes have been found using the RANSAC method by fitting plane model to 3D point clouds. Quality of the found planes mostly depends on used point tolerance value t. Usually, this value is selected empirically. In this chapter it is shown that cluster validity indices are very useful tool in measuring tolerance t from the 3D point cloud. For an assumed tolerance t the RANSAC algorithm finds the best number of planes with the quality expressed as cluster validity values. Experiments performed on synthetic and scanned 3D data have proved the benefit of cluster validity indices for validating number of planes in 3D data.

Acknowledgment This research is funded by the European Social Fund under the project "Microsensors, microactuators and controllers for mechatronic systems (Go-Smart)" (Agreement No VP1-3.1-ŠMM-08-K-01-015).

References

Asus. http://www.asus.com/Multimedia/Xtion_PRO_LIVE/. Accessed 28 Jan 2013

Ansari Z, Azeem MF, Ahmed W, Babu AV (2011) Quantitative evaluation of performance and validity indices for clustering the web navigational sessions. World Comput Sci Inf Technol J 1:217–226

Davies DL, Bouldin DW (1979) A cluster separation measure. Pattern Anal Mach Intell 1:224–227

Dunn JC (1974) Well separated clusters and optimal fuzzy partitions. J Cybern 4(1):95–104

Fischler MA, Bolles RC (1981) Random sample consensus: a paradigm for model fitting with applications to image analysis and automated cartography. Commun ACM 24(6):381–395

Gokturk SB, Yalcin H, Bamji C (2004) A time-of-flight depth sensor—system description, issues and solutions. In: Proceedings of computer vision and pattern recognition workshop, p 35

Halkidi M, Batistakis Y, Vazirgiannis M (2001) On clustering validation techniques. J Intell Inf Syst 17:107–145

Huang H, Brenner C (2011) Rule-based roof plane detection and segmentation from laser point clouds. In: Proceedings of urban remote sensing event, pp 293–296

Junhao X, Jianhua Z, Jianwei Z, Houxiang Z, Hildre H. P (2011) Fast plane detection for SLAM from noisy range images in both structured and unstructured environments. In: Proceedings of international conference on mechatronics and automation, pp 1768–1773

Kaufman L, Rousseeuw PJ (1990) Finding groups in data: an introduction to cluster analysis. Wiley, Hoboken

Microsoft, Microsoft Kinect. http://www.xbox.com/en-us/kinect. Accessed 28 Jan 2013

Milligan WG (1996) Clustering validation: results and implications for applied analyses. Clustering Classif 1:341–375

Pathak K, Birk A, Vaskevicius N, Poppinga J (2010a) Fast registration based on noisy planes with unknown correspondences for 3D mapping. IEEE Trans Rob 26(3):424–441

Pathak K, Birk A, Vaskevicius N, Pfingsthorn M, Schwertfeger S, Poppinga J (2010) Online 3D SLAM by registration of large planar surface segments and closed form pose-graph relaxation. J Field Rob 27:52–84 (Special Issue on 3D Mapping)

Rousseew PJ (1987) Silhouettes: a graphical aid to the interpretation and validation of cluster analysis. J Comput Appl Math 20:53–65

Schnabel R, Wahl R, Klein R (2007) Efficient RANSAC for point-cloud shape detection. Comput Graph Forum 26:214–226

Tarsha-Kurdi F, Landes T, Grussenmeyer P (2008) Hough-transform and extended ransac algorithms for automatic detection of 3d building roof planes from lidar data. Photogram J Finland 21(1):97–109

Turk G, Levoy M (1994) Stanford University. http://graphics.stanford.edu/data/3Dscanrep/. Accessed 29 Jan 2013

Vosselman G (2009) Advanced point cloud processing. Photogrammetric Week, Stuttgart

Feature Selection Using Adaboost for Phoneme Recognition

R. Amami, D. B. Ayed and N. Ellouze

Abstract The propose to improve a Support Vector Machines (SVM) learning accuracy by using a Real Adaboost algorithm for selecting features is presented in the chapter. This technique aims to minimize the recognition error rates and the computational effort. Hence, the Real Adaboost will be used not as classifier but as a technique for selecting features in order to keep only the relevant features that will be used to improve our systems accuracy. Since the Real Adaboost is only used for binary classifications problems, we investigate different ways of combining selected features applied to a multi-class classification task. To experiment this selection, we use the phoneme datasets from TIMIT corpus [Massachusetts Institute of Technology (MIT), SRI International and Texas Instruments, Inc. (TI)] and Mel-Frequency Cepstral Coefficients (MFCC) feature representations. It must be pointed out that before using the Real Adaboost the multi-class phoneme recognition problem should be converted into a binary one.

1 Introduction

Current studies in an automatic speech recognition (ASR) area are devoted to improving an accuracy of the ASR system in terms of error rates, memory and computation effort in both, learning and test stage. Recently, the Boosting have been proved to be an efficient technique for improving the generalized performance of different single classifiers (Freund and Schapire 1995). The Adaboost has been widely applied to face detection (Lienhart and Maydt 2002), pattern detection (Michael and Viola 2003). The aim idea of this technique is to learn an ensemble of weak classifiers which will be combined to form a single strong classifier.

R. Amami (✉) · D. B. Ayed · N. Ellouze
Department of Electrical Engineering, National School of Engineering of Tunis, University of Tunis—El Manar, Tunis, Tunisia
e-mail: Rimah.amami@yahoo.fr

Z. S. Hippe et al. (eds.), *Issues and Challenges in Artificial Intelligence*, Studies in Computational Intelligence 559, DOI: 10.1007/978-3-319-06883-1_4,

The Real Adaboost has been successfully used for feature selection in order to achieve a lower training error rate by eliminating non-effective features. Thus, the problem of selecting features in phoneme recognition filed is raised in this chapter for two main reasons; the first reason is to reduce the initial set of features originally extracted in order to reduce the classifier computational effort in the test stage. The second reason consist on including only the relevant features which are detrimental for most machine learning algorithms, and so increase the predictive performance.

The features set to be employed in the recognition task are prepared in two training-stage steps. In the first step, a feature extraction algorithm produces initial set of features (possibly large). In the second step, the initial set is reduced using a selection procedure.

The proposed selection procedure is based on the AdaBoost selection framework. This technique has been adapted in many classification and detection tasks, e.g. Viola and Jones (2001), Torralba et al. (2004) or Li et al. (2006).

In this chapter we propose a feature selection technique based on Genuine-phone and Imposter-phone which convert the multi-class problem into binary one. The remaining parts of this chapter are organized as follows: In Sect. 2 the main idea of Real Adaboost is introduced. In Sect. 3, the architecture of feature selector system is proposed. An experimental setup and results are described in Sects. 4 and 5. The conclusion is made in Sect. 6.

2 Adaboost Background

The Adaptive Boosting (AdaBoost), an iterative algorithm, was originally introduced by Feund and Schapire in 1995 (Freund and Schapire 1995, 1997). If the applied learning algorithm had a low performance, Adaboost algorithm generates a sequentially weighted set of weak classifiers in aim to create new classifiers which are more strong and operational on the training data. Hence, the AdaBoost algorithm multiple iteratively classifiers to improve the classification accuracies of many different data sets compared to the given best individual classifier. The main idea of the Adaoost is to run repeatedly a given weak learning algorithm for different probability distributions, W, over the training data. In the meantime, the Adaboost calls a Weak Learner algorithm repeatedly in a series of cycles T. Then, it assigns higher weights to the misclassified samples by the current component classifier (at cycle t), in the hopes that the new weak classifier can reduce the classification error by focusing on it. Meanwhile, lower weights will be assigned to the correctly classified samples. Thereafter, the distribution W is updated after each cycle. In the end, hypothesis produced by the weak learner from each cycle are combined into a single "Strong" hypothesis f (Meir and Ratsch 2003).

In fact, the important theoretical property of the Adaboost is that if the component classifiers consistently have accuracy slightly better than half, then the training error of the final hypothesis drops to zero exponentially fast. This means that the component classifiers need to be slightly better than random estimation (Li et al. 2008).

The Real AdaBoost is the generalization of the basic AdaBoost algorithm introduced by Fruend and Schapire (Vezhnevets and Vezhnevets 2005). Thus, it should be treated as a basic fundamental Boosting algorithm.

The particularity of this algorithm is Confidence-rated Predictions function that is a map from a sample space X to a real-valued space R instead of Boolean prediction. We are given a set of training samples $(x_1, c_1),\ldots,(x_n, c_n)$ where x_i is the input, and each output c_i belong to $\{1,-1\}$ is a class label. Let h(x) denotes the weak classifier and the confidence of $h_f(x)$ is defined as (Schapire and Singer 1999):

$$conf\ h_f(x) = \left| \sum_{t=1}^{T} h_t(x) - b \right| \tag{1}$$

where b is a threshold of default value equal to zero.

3 SVM Background

The Support Vector Machine (SVM) is a learning algorithm widely applied for pattern recognition and regression problems based on the Structural Risk Minimization (SRM) induction principle (Cortes and Vapnik 1995). SVM maximizes a margin which is a distance from a separating hyperplane to the closest positive or negative sample between classes. A subset of training samples is chosen as support vectors. They determine the decision boundary hyperplane of the classifier.

Applying a kernel trick that maps an input vector into a higher dimensional feature space, allows the SVM for approximating a non-linear function (Cortes and Vapnik 1995; Amami et al. 2012). In this chapter, we propose to use the radial basis function kernel (RBF).

4 Feature Selection System

The architecture of our phoneme recognition system is described in this section. The proposed system aims to select the relevant features using the Real Adaboost. Its architecture consists of: (1) conversion from speech waveform to spectrogram, (2) spectrogram to Mel-Frequency Cepstral Coefficients (MFCC) spectrum using the spectral analysis, (3) segmentation of the phoneme data sets to 7 sub-phoneme data sets, (4) selection of significant MFCC features using the Real Adaboost, (5) combination of all significant selected features per each phoneme (the Real Adaboost is a algorithm for binary classification task thus, it is a necessary step), (6) initiate the phoneme recognition task using the SVM, see Fig. 1.

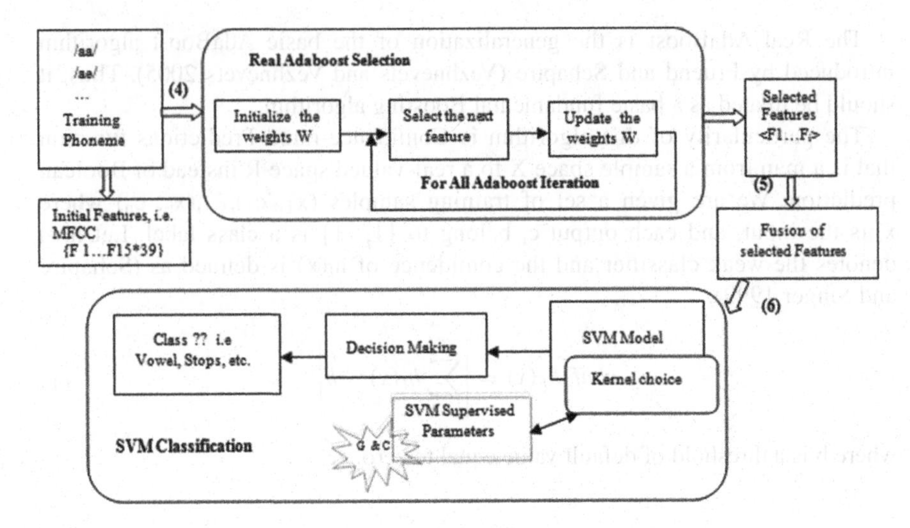

Fig. 1 Architecture of the feature selection based the Adaboost system

The Real Adaboost is applied only to a binary classifications task. However, we propose another way to apply this algorithm to multi-class problem phoneme recognition by using the "dual difference class", i.e., Genuine and Imposter. The Genuine and Imposter strategy is used to transform the multi-class problem into a binary one. Since we consider the Real Adaboost as a feature selection tool instead of a classifier training procedure, we will get, finally, features selected and not the set of classifiers. Then, the feature selected can be processed further by SVM to build classifier and conduct the recognition task. In order to apply the Real Adaboost for phoneme recognition, we should convert the multi-class problem into a binary one. One-vs-one and one-vs-rest are two typical ways. In this chapter, we adopt the one-vs-rest technique. Given two phoneme samples from the training set, if they are from the one same class (i.e. Vowel, Affricate, Nasal, etc.), they will be put into the genuine class (positive label), otherwise into the imposter class (negative label).

5 Experimental Setup

As discussed in the previous section, the first step in an ASR system is the feature extraction. It converts the speech waveform to the set of parametric representation. Hence, we have used the MFCC feature extractor (Davis and Mermelstein 1980). The proposed approach has been evaluated using the dialect DR1 (New England) from TIMIT corpus (Garofolo et al. 1993).

For the nonlinear SVM approach, we choose the RBF (Gaussian) kernel trick, this choice has been made after a previous study done on our data sets with different kernel tricks (Linear, Polynomial, Sigmoid) (Amami et al. 2013a).

In the current work, we use the "one-against-one" method and the voting strategy. As the classification performance of SVMs is mainly affected by its model

Table 1 Binary phoneme recognition rates using two techniques: SVM with all features and SVM using the selected features

Phoneme	Techniques		#Features	
	SVM-RBF (%)	SVM-RA (%)	Train	Test
Vowel versus all	87	86	63	75
SemiVowel versus all	89	88	67	68
Stops versus all	79	41	72	74
Others versus all	95	93	79	66
Nasal versus all	94	92	67	65
Fricative versus all	94	87	72	68
Affricate versus all	99	99	70	62

parameters particularly the Gaussian width and the regularization parameter C, we set, for all experiments, gamma as a value within 1/K where K is the features dimension and C as value within 10. Those parameters are the suitable for SVM on our data sets. Furthermore, the input speech signal is segmented into frames of 16 ms with a Hamming window. Moreover, each phoneme has a feature vector which contains 39 MFCC coefficients including first delta (Delta), second delta (Delta-Delta) and the energy.

We have used Genuine/Imposter strategy which consists of selecting the significant features for a phoneme X against the rest of phonemes. Since, we are dealing with a multi-class recognition task; two techniques have been applying in order to combine the selected features for each phoneme into one macro-class database. It will be used as input to SVM for the recognition task.

6 Results

In the most available sounds databases, the utterance does not capture all possible variations for each sample due to noise, vocabulary size, confusability, etc. Feature extraction and selection is one of the solutions to this problem. This technique can reduce the dimensionality of phoneme in the feature space. Hence, we introduce the Real Adaboost as a feature selection tool. Obviously, a smaller set of features implies less computational effort in the test stage since the computation of less relevant features will be removed. A successful selection method might also improve recognition accuracy and enhance robustness. Comparisons are made between the Adaboost method and also the non-boosted classifier.

The selection procedure returns the T features with highest evaluation score in respect to the current weights of W. This selection approach tends to produce redundant selections sets, since the features with similar evaluation scores may exhibit a high degree of interdependence. This phenomenon is discussed by Cover (1974). As it is seen in the Table 1, for a problem described by 117 features the dimension of the selected features vector ranges from 62 to 79 features for both, the train and the test stages. Meanwhile, the recognition system rates with SVM-RBF outperform slightly the recognition system with the proposed SVM-RA. The highest performance is achieved with the Affricate phones within 99 % of correct classification.

Table 2 Multi-class phoneme recognition rates using three strategies: single SVM, SVM-MCU and SVM-MCC

Strategy	Rec. (%)	Runtime (s)	#Features	
SVM-MCU	49	457	490	478
SVM-MCC	57	164	117	115
SVM-RBF	74	176	117	117

It must be pointed out that the runtime of the proposed technique SVM-RA is better than the SVM-RBF since the dimension of the features vector was reduced.

For the multi-class recognition task, the results are from two sets of experiments:

- SVM-MCU: this technique is based on a macro-class database including all samples and the features selected for each phoneme data set and even the redundant selections sets.
- SVM-MCC: this technique is based on a macro-class database including all samples and the features selected per each phoneme but excluding the redundant selections sets.

Before the feature selection, the original vector dimension of each phoneme was fixed to 117 (3 middle-frames) and the number of iteration of the Real Adaboost was fixed to 50. According to the experimental results presented in the Table 2, the SVM-RBF outperforms the SVM based on both features selected techniques (SVM-MCU and SVM-MCC). We can, also, learn that classifier based on original features performs better than the classifier based on feature selection. This result can be explained by the fact that the initial dimension of each feature is very low (117), thus, the feature selection using the Real Adaboost failed.

In the last decades, some experiments have been carried out in order to better understand why boosting sometimes leads to deterioration in generalization of performance (Amami et al. 2013b). Freund and Schapire put this down to overfitting a large number of trials T (T stands for the Adaboost iteration number) which allows to composite classifier to become very complex (Freund and Schapire 1997; Quinlan 1996).

Besides these, boosting tries to build a strong classifier from weak classifiers. However, if the performance of the classifier is better than random estimation (i.e. SVM), then boosting may not result in a strong classifier and this method would be going against the gain of the Boosting principle and do not achieve the desired results.

7 Conclusion

In our research, we have used the Adaboost as the feature selector. The feature selection via the Real Adaboost has been also studied with the aim of reducing the number of irrelevant features. It can be seen, from the results gained, that the recognition system utilizing the Real Adaboost feature selection method has not improved the recognition accuracy in comparison to the SVM-RBF.

The future work will be centered on extending our approach in order to improve the phoneme recognition accuracy.

References

Amami R, Ben Ayed D, Ellouze N (2012) Phoneme recognition using support vector machine and different features representations. Adv Intell Soft Comput 151:587–595

Amami R, Ben Ayed D, Ellouze N (2013a) Practical selection of SVM supervised parameters with different feature representations for vowel recognition. Int J Digit Content Technol Appl 7(9):418–424

Amami R, Ben Ayed D, Ellouze N (2013b) Adaboost with SVM using GMM supervector for imbalanced phoneme data. In: Proceedings of 6th International IEEE conference on human system interaction, pp 328–333

Cortes C, Vapnik V (1995) Support-vector networks. Mach Learn 20:273–297

Cover T (1974) The best two independent measurements are not the two best. IEEE Trans Syst Man Cybern 4:116–117

Davis SB, Mermelstein P (1980) Comparison of parametric representations for monosyllabic word recognition in continuously spoken sentences. IEEE Trans Acoust, Speech, Signal Proc 28:357–366

Freund Y, Schapire RE (1997) A short introduction to boosting. J Japan Soc Artif Intell 14(5):771–780

Freund Y, Schapire RE (1995) A decision-theoretic generalization of online learning and an application to boosting. In: Proceedings of 2nd European conference on computational learning theory, pp 23–37

Garofolo JS, Lamel LF, Fisher WM, Fiscus JG, Pallett DS, Dahlgren NL, Zue V (1993) Timit acoustic-phonetic continuous speech corpus. Texas instruments and massachusetts institute of technology

Li F, Košecká J, Wechsler H (2006) Strangeness based feature selection for part based recognition. In: IEEE conference on computer vision and pattern recognition

Li X, Wang L, Sung E (2008) Adaboost with SVM-based compnent classifers. Eng Appl Artif Intell 21:785–795

Lienhart R, Maydt J (2002) An extended set of Haar-like features for rapid object detection. In: Proceedings of IEEE conference on image processing, pp 900–903

Michael J, Viola P (2003) Face recognition using boosted local features. In: Proceedings of international conference on computer vision, pp 1–8

Meir R, Ratsch G (2003) An introduction to boosting and leveraging. Advanced lectures on machine learning, Springer, Berlin, pp 119–184

Schapire RE, Singer Y (1999) Improved boosting algorithms using confidence-rated predictions. Mach Learn 37(3):297–336

Quinlan JR (1996) Bagging, boosting, and c4.5. In: Proceedings of 13th national conference on artificial intelligence, pp 725–730

Torralba A, Murphy K, Freeman W (2004) Sharing features: efficient boosting procedures for multiclass object detection. In: IEEE conference on computer vision and pattern recognition, pp 762–769

Vezhnevets A, Vezhnevets V (2005) Modest adaboost—teaching adaboost to generalize better. Graphicon, pp 322–325

Viola P, Jones M (2001) Robust real-time face detection. J Comput Vision 2:137–154

Competence of Case-Based Reasoning System Utilizing a Soft Density-Based Spatial Clustering of Application with Noise Method

A. Smiti and Z. Elouedi

Abstract Clustering is one of the most valuable methods of computational intelligence field, in particular, in human–Computer Systems Interaction context, in which sets of related objects are cataloged into clusters. In this background, we put a spotlight on the importance of the clustering exploit in the competence computing for Case Based Reasoning (CBR) systems. For that, we apply an efficient clustering technique, named "Soft DBSCAN" that combines Density-Based Clustering of Application with Noise (DBSCAN) and fuzzy set theory, on competence model. Our clustering method is galvanized by Fuzzy C Means in the way of using the fuzzy membership functions. The results of our method show that it is efficient not only in handling noises, contrary to Fuzzy C Means, but also, able to assign one data point into more than one cluster, and in particular it shows high accuracy for predicting the competence of CBR. Simulative experiments are carried out on a variety of datasets, throughout different evaluation's criteria, which emphasize the soft DBSCAN's success and cluster validity to check the good quality of clustering results and its usefulness in the competence of the CBR.

1 Introduction

Cluster analysis has an imperative responsibility in analysis of the human-computer systems interaction activities. The leading intention of clustering is to simplify statistical study by assembling equivalent objects into clusters that in some sense belong together because of correlated characteristics. A colossal number of clustering

---

A. Smiti (✉) · Z. Elouedi
Larodec Institut Supérieur de Gestion de Tunis, University of Tunis, Tunis, Tunisia
e-mail: smiti.abir@gmail.com

Z. Elouedi
e-mail: zied.elouedi@gmx.fr

Z. S. Hippe et al. (eds.), *Issues and Challenges in Artificial Intelligence*,
Studies in Computational Intelligence 559, DOI: 10.1007/978-3-319-06883-1_5,

algorithms can be found in the literature. They can be classified as partitioning, hierarchical, density (or neighborhood)-based and grid-based methods. One of the most well known admired and partitioned based clustering algorithms is KMeans (MacQueen 1967), its formula trails a simple manner to organize a given data set through a certain number of clusters (assume k clusters) predetermined a priori. From the density based clustering, the most extensively used method is DBSCAN (Density-Based Spatial Clustering of Application with Noise) (Ester et al. 1996). It is designed to ascertain clusters of random shape as well as to differentiate noises. From another point of view, the clustering methods can be investigated whether they are crisp or fuzzy clustering methods. The straight clustering methods check that each point of the data set belongs to exactly one cluster. Fuzzy set theory proposed in Zadeh (1965) gave an idea of uncertainty of belonging which was described by a membership function. The employ of fuzzy sets recommends imprecise class membership information. One branch of fuzzy clustering research has based on fuzzy relation, like fuzzy equivalent relation-based hierarchical clustering (Klir and Yuan 1995) and FN-DBSCAN (Parker and Kandel 2010). Another branch of fuzzy clustering has focused on objective functions, as Fuzzy C Means (FCM) (Han 2005) and competitive agglomeration algorithm (Frigui and Krishnapuram 1996). Most of the proposed fuzzy clustering methods are based on the Fuzzy C Means algorithm (Han 2005). These methods consider the fuzziness of clustering as being assigned to some clusters with certain degrees of membership, which is not the case for density-based method. Hence, in order to make the density-based clustering algorithms robust, extending them with the fuzzy set theory has attracted our attentions where we have proposed a novel clustering algorithm called "Soft DBSCAN" in Smiti and Elouedi (2013a). Our idea is to improve the clusters generated by DBSCAN by fuzzy set theory which is based on an objective function, in order to produce optimal fuzzy partitions. Our soft method provides a similar result as Fuzzy C Means, but it is simple and superior in handling outlier points.

In this chapter, we underline the significance of our Soft clustering method in the computing of the competence for the CBR system. For that, we will apply our Soft DBSCAN to the model proposed by (M&S) (Smyth and McKenna 2001) to show its performance. The cheering results achieved on some data sets are revealed and discussed. The rest of this chapter is organized as follows. In the next Section, CBR system and the concept of competence model will be presented. Section 2 describes our new soft DBSCAN clustering method. Section 3 presents the application of our soft clustering technique on the Smyth's model. Experimental setup and results are given in Sect. 4. The chapter concludes in Sect. 5.

2 Case Based Reasoning systems

One of the huge intentions of human and computer systems interaction area is to create smart systems able to recognize and follow human logic. Among these systems, CBR is a diversity of reasoning by analogy and it is able to discover a

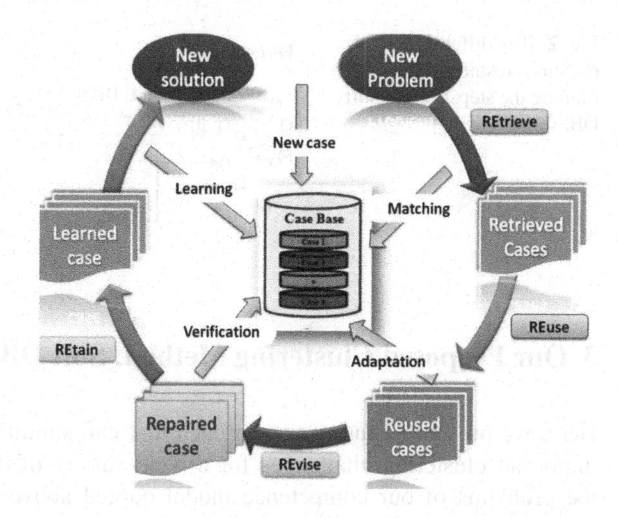

Fig. 1 The four REs CBR cycle (REtrieve, REuse, REvise, REtain)

solution to a problem by employing its luggage of knowledge or experiences which are presented in form of cases. To solve the problems, CBR system calls the past cases, it reminds to the similar situations already met. Then, it compares them with the current situation to build a new solution which, in turn, will be incorporated it into the existing case base (CB). Different ways have been recommended to illustrate the CBR process, but the traditional the four REs CBR cycle (REtrieve, REuse, REvise, REtain) is the most recurrently used: The new problem is matched against cases in the case base and one or more related cases are retrieved. A solution advocated by the matching cases is then reused and tested for success. Unless the retrieved case is a close match, the solution will have to be revised generating a new case that can be maintained (See Fig. 1).

Actually, the performance of CBR can be measured according to a major criterion: Competence or named coverage which is the range of target problems that can be successfully solved (Smiti and Elouedi 2013b). However, it is complicated to count the competence of the system, for the motive that the precise nature of the rapport between the case base and competence is versatile and not well understood. Accordingly, we entail a theoretical model that authorizes the competence of a case base to be guess or estimated. In the literature, few models have been proposed to represent the coverage of the CB, we can mention the competence model (M&S) proposed in Smyth and McKenna (2001), which is the most used model in CBR and it is the origin of many maintenance works including ours (Smiti and Elouedi 2012b). It is assuming that the competence of the CB is simply the sum of the coverage of each group of similar cases. This study has drawn attention to the enormity of modeling CBR competence. Conversely, it suffers from some shortages for instance it does not pay attention to the condition of non-uniform distributed case-bases. Besides, it is delicate to erroneous cases like the noises. Further, it does not touch the challenges posed by the compilation of natural data which is often vague. To elevate these troubles, we need an efficient clustering method which while creating competence groups, can also handle the problems cited above.

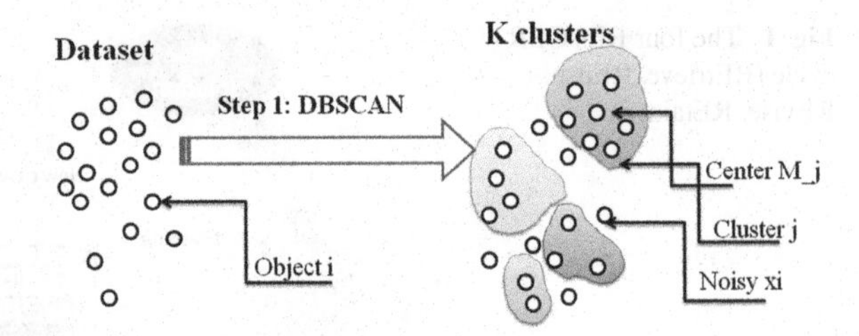

Fig. 2 The different partitions resulting from running the step 1 of our soft DBSCAN

3 Our Proposed Clustering Method: Soft DBSCAN

Here, we present a clustering approach that can simultaneously address several important clustering challenges for a wide variety of data sets and can handle the problems of our competence model quoted above. In particular, our algorithm can manage instances expected to be noisy, it can create clusters with different shapes, and it allows the elements to have a degree of membership for each cluster.

In this section, we present a new fuzzy clustering method named "soft DBSCAN". Much of the strength of our approach comes from FCM's ideas. The plan of our "soft DBSCAN" is to make the DBSCAN's clusters robust, extending them with the fuzzy set theory. Thusly, the Soft DBSCAN's first stage runs DBSCAN which creates, many seed clusters, with a bunch of noisy points. Each noisy is consider as one cluster. These determined groups, in addition of noisy clusters, with their centers, offer a good estimate for initial degrees of membership which express proximities of data entities to the cluster centers. We update the membership values, in every iteration, since these last ones depend on the new cluster centers. When the cluster's center stabilizes, our "soft DBSCAN" algorithm stops. The basic process of the proposed method is in the following steps:

First, we run DBSCAN. The result of this step is k clusters C_k with determined centroid M_i (Where i = 1: K) and number x of noisy points.

The data were partitioned in K clusters, so several local datasets are generated and points expected to be noisy are detected. The data number of each local dataset is smaller than the initial dataset. Based on their characters, we will try to estimate initial degrees of membership values (Fig. 2).

Second step consists of estimating the number of clusters and the initial matrix of membership: As we know, the noises are objects that do not belong to any set of similar cases. For that, we consider that each point of noises is one cluster apart. As a result, the total number of clusters becomes c = k+x. Thereafter, we create the initial fuzzy partition $c \times n$ matrix $U = (u_{ij}) = (\overrightarrow{u_1}, \overrightarrow{u_2} \ldots, \overrightarrow{u_n})$, as follows:

$$u_{ij} = \begin{cases} 1 \; if & x_{ij} \in c_j \\ 0 \; otherwise. \end{cases}$$

In the third step, we compute the jth center V_j which is given by:

$$v_j = \frac{\sum_{j=1}^{n} \mu_{ij}^{m} x_{ij}}{\sum_{j=1}^{n} \mu_{ij}^{m}}; \text{(m > 1)} \tag{1}$$

In the fourth step, we compute the values for the distance (D_{ij}) from the sample x_{ij}, of the center V_j, of the jth class. For that, we have to employ distance which takes into account the arbitrary shape of the clusters.

As we have mentioned, our goal is to create non uniform clusters. Or, the reason that FCM and many other clustering techniques can only work well for spherical shaped clusters, is in the objective function the distances between data points to the centers of the clusters are calculated by Euclidian distances. The best technique to overcome the above drawback is Mahalanobis distance (MD) (Filzmoser et al. 2005) because it takes into consideration the covariance among the variables in calculating distances. With this measure, the problems of scale and correlation inherent in the other distance such as Euclidean one are no longer an issue. Hence, MD is efficient for the non uniform distribution and arbitrarily shaped clusters because it deals with clusters of different densities and silhouettes.

Given p-dimensional multivariate sample (cases) x_i (i = 1, 2..., n), the Mahalanobis distance is defined as:

$$MD_i = ((x_i - t)^T C_n^{-1} (x_i - t)^{1/2} \tag{2}$$

where t is the estimated multivariate location and C_n the estimated covariance matrix:

$$C_n = \frac{1}{n-1} \sum_{i=1}^{n} (x_i - \overline{X}_n)(x_i - \overline{X}_n)^T \tag{3}$$

where $\overline{X}_n$the mean of the cluster and n is is the number of cases in this cluster. With this distance, we can update our membership values μ_{ik} for all i:

$$\mu_{ik} = \left[\sum_{j=1}^{c} \left(\frac{MD_{ik}}{MD_{jk}} \right)^{\frac{2}{m-1}} \right]^{-1} \tag{4}$$

where MD_{ik} is the Mahalanobis distance between x_k and v_k.

If the difference between the actual membership values μ_{ik}^{actual} and the ones before the update μ_{ik}^{prec} is superior to the tolerance level ξ, we stop our algorithm, else we return to the third step:

$$||\mu^{actual} - \mu^{prec}|| \leq \xi \begin{cases} \textit{Stop} \\ \textit{Return to step } 3, \ \textit{otherwise}. \end{cases}$$

In this way, our method generates different clusters c_j, with v_{ij} centers. The sets which contain only one instance are considered as noises.

Subsequently, we obtain a new technique which has a number of good aspects: it can create regions which may have an arbitrary shape and the points inside a region may be arbitrarily distributed, it can detect points expected to be noises and it is able to assign one data point into more than one cluster by affecting to each observation a "degree of membership" to each of the classes in a way that is consistent with the distribution of the data.

4 Application of Our Soft DBSCAN in the Competence of CBR

In this Section, we propose a competence for CBR strategy method based on the M&S model, and we adapt in this study our new "Soft DBSCAN" described in the previous Section, to perform the creation of the competence groups. This model is assuming that the competence is based on a number of factors including the size and the density of cases in the case base. The number and density of cases can be readily measured. Actually, the individual competence contribution of a single case within a dense collection will be lower than the contribution of the same case within a sparse group; dense groups contain greater redundancy than sparse groups. The density of an individual case (Dens) can be defined as the average similarity between this case (c) and other clusters of cases called competence groups (Eq. 5). Hence, the density of a cluster of cases is measured as a whole as the average local density over all cases in the group (G) (Eq. 6). The coverage of each competence group (Cov) is then measured by the group size and density (Eq. 7). In the final step, the overall competence of the case base is simply the sum of the coverage of each group (Eq. 8).

$$Dens(c, G) = \frac{\sum_{c' \in G - c} Sim(c, c')}{|G - 1|} \tag{5}$$

$$Dens(G) = \frac{\sum_{c \in G} Dens(c, G)}{|G|} \tag{6}$$

$$Cov(G) = 1 + ||G| \times (1 - Dens(G))| \tag{7}$$

$$Total\ Coverage(G) = \sum_{Gi \in G} Cov(G_i) \tag{8}$$

Obviously, by applying the idea described above, we certify that we need to create multiple, groups from the case base that are located on different sites. Each group contains cases that are closely related to each other. In that way, we can define the coverage group. So, we prefer a clustering approach that can simultaneously address several important clustering challenges for a wide variety of data sets. In particular, our chosen clustering algorithm can manage instances expected to be noisy, it can create clusters with different shapes, and it allows the elements to have a degree of membership for each cluster. To overcome all these conditions, we use our new fuzzy clustering method "soft DBSCAN" proposed in previous Section. Our soft DBSCAN is an appropriate clustering method for the M&S model. This occurs because our technique has a number

Table 1 Description of databases

Dataset	Ref.	#instances	#attributes
IRIS	IR-150	150	4
Ecoli	EC-336	336	8
IONOSPHERE	IO-351	351	34
Breast-W	BW-698	698	9
Blood-T	BT-748	748	5
Indian	IN-768	768	9

of good aspects: it can create regions which may have an arbitrary shape and the points inside a region may be arbitrarily distributed, it can detect cases expected to be noises and it is able to assign one data point into more than one cluster by affecting to each observation a "degree of membership" to each of the classes in a way that is consistent with the distribution of the data. We can say that it shows good performance comparing to other clustering algorithms, in the context of the competence computing strategy.

5 Results and Analysis

In this section, we shall use experimental results to show the clustering performance of soft DBSCAN and our competence model with this clustering technique. We experimented with ten diverse data sets, obtained from the U.C.I. repository (Asuncion and Newman 2007) (Table 1).

To test the performance of our proposed Soft DBSCAN , we will examine its quality comparing to FCM method, to prove that our method is mightier than the FCM not only in detecting noises, but also in minimizing the objective function. For FCM, we pick the value of K equal to the number of classes in the data sets. For Soft DBSCAN, the suggested values are MinPts = 4 and Eps = 0.2. We evaluate the clustering algorithms from various aspects: We analyze "the accuracy of the algorithms" (PCC). It is the mean of correct classification over stratified tenfold cross validation. We employ "the objective function" (OBJ). If our method can obtain a smaller value for this function than FCM algorithm can do, then it will be more powerful. We analyze "the Partition Coefficient" (PC) (Guill et al. 2007) to measure the amount of overlap between clusters. The larger this measure is the more density in clusters and more efficiency in clustering process. This measure is calculated as following:

$$PC = \frac{1}{n}\sum_{i=1}^{c}\sum_{i=1}^{n}\mu_{ij}^{2}; \quad \frac{1}{c} \leq PC \leq 1; \tag{9}$$

We use "the Partition Entropy" (PE) (Guill et al. 2007). The less this criterion, the better efficiency for clustering is achieved and it is calculated as following:

$$PE = -\frac{1}{n}\sum_{i=1}^{c}\sum_{i=1}^{n}\mu_{ij}\log\mu_{ij}; \quad 0 \leq PE \leq \log c; \tag{10}$$

Table 2 Comparison of classification accuracy (PCC), objective function (OBJ) and fuzzy performance index (FPI)

Ref.	PCC		OBJ		FPI	
	Soft DBSCAN	FCM	Soft DBSCAN	FCM	Soft DBSCAN	FCM
IR-150	0.98	0.66	15.94	18.12	0.93	0.93
EC-336	0.95	0.86	25.71	29.32	0.47	0.43
IO-351	0.89	0.88	133.62	133.97	0.86	0.72
BW-698	0.90	0.90	27.49	29.43	0.90	0.92
BT-748	0.90	0.85	50.21	70.58	0.67	0.91
IN-768	0.98	0.82	40.49	46.25	0.64	0.48

Table 3 Comparison of the partition coefficient (PC), the partition entropy (PE) and silhouette values (Sih)

Ref.	PC		PE		Sih	
	Soft DBSCAN	FCM	Soft DBSCAN	FCM	Soft DBSCAN	FCM
IR-150	0.35	0.34	0.56	0.55	0.88	0.86
EC-336	0.36	0.28	0.43	0.63	0.92	0.77
IO-351	0.55	0.65	0.51	0.52	0.72	0.44
BW-698	0.37	0.37	0.53	0.50	0.03	0.07
BT-748	0.77	0.41	0.35	0.46	0.38	0.01
IN-768	0.68	0.25	0.21	0.67	0.14	0.08

In addition, we examine the "Silhouette Plot" (Sih) where its value is calculated by the means similarity of each plot to its own cluster minus the mean similarity to the next most similar cluster (given by the length of the lines) with the mean in the right hand column, and the average silhouette width. Finally, we engage "Fuzzy Performance Index" (FPI). It is a measure of the degree of separation (fuzziness) between fuzzy c-partitions and the data classified through the cluster analysis (Fridgen et al. 2004). It is defined as:

$$FPI = 1 - \frac{c}{(c-1)}\left[1 - \sum_{i=1}^{c}\sum_{j=1}^{n}\mu_{ij}^{2}/n\right]; \quad 0 \leq FPI \leq 1; \tag{11}$$

From Tables 2 and 3, we observe that the results obtained using our Soft DBSCAN method is remarkably better than the one provided by the FCM policy:

As seen in the Table 2, our proposed algorithm shows uniformly high accuracy ranging 0.89–0.98 for all data sets. For instance, by applying our Soft DBSCAN to the "Indian (IN-768)" database, the PCC increases from 0.82 for FCM to 0.98. In addition, it reaches 0.98 PCC which presents a great difference compared to the one given by FCM: just 0.66 of PCC, for the dataset "Iris (IR-150)". These good values are elucidated by handling noisy objects by our Soft DBSCAN method.

Table 4 Comparing (M&S) model to the one with "soft DBSCAN"

Case	M&S	M&S with soft DBSCAN
IR-150	4.01	3.87
EC-336	9.24	9.01
BW-698	7.25	5.89
BT-748	17.90	11.32
IN-768	6.10	4.91

The same observations are made by the objective function, where the values provided by our method show better values than FCM. In addition, the results of FPI in the most datasets show that soft DBSCAN yields the lowest values of the objective function and the highest values of FPI comparing to FCM, especially for "Ecoli (EC-336)" dataset, where our method generates only 50.21 as OBJ and 0.47 as FPI, whereas 70.58 and 0.43 respectively by FCM, for the same dataset.

From Table 3, the proposed algorithm shows uniformly high PC with mean of 0.513 and low PE with mean of 0.431. However, FCM algorithm produces only 0.383 of PC and 0.555 of PE. This result is explained by the large density in clusters created by Soft DBSCAN, and this is due to the agreeable membership values provided by our method, so, the best partition is achieved. For example, for the dataset "Blood-T (BT-748)", Soft DBSCAN outperforms FCM for PC criterion by the value of 0.77, whereas for FCM is only 0.41. As for PE criterion, Soft DBSCAN and FCM averaged out 0.35 and 0.46, respectively.

Although, it is obvious from the clustering results that our method offers positive results, the silhouette values shown in Table 3 clearly indicate that Soft DBSCAN is better than FCM. Based on this evaluation criterion, the samples of the dataset "Ionosphere (IO-351)" have their silhouette value larger than 0.72 when using soft DBSCAN, compared with less than 0.44 for FCM. The same conclusion can be drawn for the other datasets.

Finally, we have to mention that Soft DBSCAN's run time, required to build the clusters, is nearly equal to the two times the run time of FCM. On average to build the clusters, our algorithm took 10.21 seconds, per contra; FCM took 4.59 s, since our policy uses DBSCAN method and calculates the membership values.

In the second part of our experimentation, we test the performance of our competence model with "Soft DBSCAN". For that, we will compare it with the well-known competence model (M&S), using the same benchmark data sets as described above. For this comparison, we use Percentage error as evaluated index (Eq. 12), which represents the relative error of coverage computed by using the (M&S), and the one with "Soft DBSCAN".

$$Error(\%) = \frac{|EstimateComp - PCC|}{PCC} * 100 \tag{12}$$

Table 4 proves optimistic results for our competence model. The Percentage error of M&S model with our Soft DBSCAN is pretty inferior to the one of M&S model. This is due to the application of our Soft DBSCAN.

6 Conclusions

In this chapter, we have applied our proposed Soft DBSCAN on the competence model of Case based reasoning system. It is an improvement of DBSCAN density clustering and combined to fuzzy set theory in terms of solving its disadvantages. Our proposed method does not only outperform FCM clustering by detecting points expected to be noises and handling the arbitrary shape, but also by generating more dense clusters. Results of experimentations have shown interesting results for the competence model with Soft DBSCAN as clustering technique. Future tasks consist of applying this model in the maintenance of the case based reasoning systems.

References

Asuncion A, Newman D (2007) UCI machine learning repository. http://www.ics.uci.edu/mlearn

Ester M, Kriegel HP, Sander J, Xu X (1996) A density-based algorithm for discovering clusters in large spatial databases with noise. In: Proceedings of 2nd international conference on knowledge discovery, pp 226–231

Filzmoser P, Garrett RG, Reimann C (2005) Multivariate outlier detection in exploration geochemistry. Comput Geosci 31:579–587

Fridgen JJ, Kitchen NR, Sudduth KA, Drummond ST, Wiebold WJ, Fraisse CW (2004) Management zone analyst (MZA): software for subfield management zone delineation. Agron J 96:100–108

Frigui H, Krishnapuram R (1996) A robust clustering algorithm based on competitive agglomeration and soft rejection of outliers. In: IEEE computer society, pp 550–555

Guill A, Gonzalez J, Rojas I, Pomares H, Herrera LJ, Valenzuela O, Prieto A (2007) Using fuzzy logic to improve a clustering technique for function approximation. Neurocomputing 70(16–18):2853–2860

Han J (2005) Data mining concepts and techniques. Morgan Kaufmann Publishers Inc, San Francisco, CA

Klir GJ, Yuan B (1995) Fuzzy sets and fuzzy logic theory and applications. Prentice Hall PTR, Upper Saddle River

MacQueen JB (1967) Some methods for classification and analysis of multivariate observations. In Proceedings of 5th Berkeley symposium on mathematical statistics and probability, vol 1, pp 281–297

Parker L, Kandel A (2010) Scalable fuzzy neighborhood dbscan. In: FUZZ-IEEE, pp 1–8

Smiti A, Elouedi Z (2012a) Dbscan-gm: an improved clustering method based on gaussian means and dbscan techniques. In: Proceedings of international conference on intelligent engineering systems, pp 573–578

Smiti A, Elouedi Z (2012b) Competence and performance-improving approach for maintaining case-based reasoning systems. In: Proceedings of international conference on computational intelligence and information technology, pp 356–361

Smiti A, Elouedi Z (2013a) Soft DBSCAN: improving DBSCAN clustering method using fuzzy set theory. In: Proceedings of 6th international conference on human system interaction, pp 380–385

Smiti A, Elouedi Z (2013b) Modeling competence for case based reasoning systems using clustering. In: Proceedings of 26th international conference Florida artificial intelligence research society, pp 399–404

Smyth B, McKenna E (2001) Competence models and the maintenance problem. Comput Intell 17(2):235–249

Zadeh L (1965) Fuzzy sets. Inf Control 8(3):338–353

Intelligent Information System Based on Logic of Plausible Reasoning

B. Śnieżyński, S. Kluska-Nawarecka, E. Nawarecki and D. Wilk-Kołodziejczyk

Abstract The chapter presents a methodology for the application of a formalism of the Logic of Plausible Reasoning (LPR) to create knowledge about a specific problem area. In this case, the methodology has been related to the task of obtaining information about the innovative casting technologies. In the search for documents, LPR gives a much greater expressive power than the commonly used keywords. The discussion is illustrated with the results obtained using a pilot version of the original information tool. It also presents a description of intelligent information system based on the LPR and the results of tests on the functionality and performance parameters of the system.

1 Introduction

Recently, great interest has aroused various issues related to acquiring and processing of information and domain knowledge contained in large data sets, such as web services, data warehouses, or subject-specific databases.

Although numerous studies have recently appeared related to this problem area, there is still a number of issues to be addressed concerning, among others, scalability

B. Śnieżyński (✉) · E. Nawarecki · D. Wilk-Kołodziejczyk
AGH University of Science and Technology, Kraków, Poland
e-mail: bartlomiej.sniezynski@agh.edu.pl

E. Nawarecki
e-mail: edward.nawarecki@agh.edu.pl

D. Wilk-Kołodziejczyk
e-mail: dorota.wilk@iod.krakow.pl

S. Kluska-Nawarecka · D. Wilk-Kołodziejczyk
Foundry Research Institute in Cracow, Kraków, Poland

S. Kluska-Nawarecka · E. Nawarecki
University of Computer Science and Skills, Łódź, Poland

Z. S. Hippe et al. (eds.), *Issues and Challenges in Artificial Intelligence*, Studies in Computational Intelligence 559, DOI: 10.1007/978-3-319-06883-1_6,

(e.g. of the Internet), search and classification (especially for the information expressed in linguistic form), processing and representation of the acquired knowledge (bringing it to the form easy to be interpreted by humans).

This study shows the concept and the results of a pilot implementation of an information system that uses the possibilities offered by the LPR formalism. This idea was based on the observation that in the intention of the authors of LPR (Collins and Michalski 1989) this logic should correspond to the perception of the world in a manner used by man and, consequently, its understanding becomes in certain situations easier and more efficient.

In the original approach (Collins and Michalski 1989), the LPR formalism is not adapted to the implementation of automatic reasoning system and, consequently, some of its elements are redundant, while other are lacking. Therefore, the chapter proposes a modification of the LPR formalism, defined as LPR^0, facilitating its use in the problem area outlined above. The second part describes the implemented solution, which is an intelligent information system dedicated to the search and processing of domain knowledge based on the analysis of the contents of the source documents. Operation of the system is illustrated by the example of searching for information (knowledge) of innovative technologies in the foundry industry. Preliminary tests of functionality and performance is also carried out, allowing the measurement of the efficiency of the proposed solution.

2 Related Research

LPR was proposed by Alan Collins and Richard Michalski, who in 1989 published their article entitled "The Logic of Plausible Reasoning, A Core Theory" (Collins and Michalski 1989). The aim of this study was to identify patterns of reasoning used by humans and create a formal system based on the variable-valued logic calculus (Michalski 1983), which would allow for the representation of these patterns. The basic operations performed on the knowledge represented in the LPR include:

abduction and deduction—are used to explain and predict the characteristics of objects based on domain knowledge;

generalisation and specialisation—allow for generalisation and refining of information by changing the set of objects to which this information relates to a set larger or smaller;

abstraction and concretisation—change the level of detail in description of objects;

similarity and contrast—allow the inference by analogy or lack of similarity between objects.

The experimental results confirming that the methods of reasoning used by humans can be represented in the LPR are presented in subsequent paper (Cawsey 1991). The objective set by the creators has caused that LPR is significantly different from other known methods of knowledge representation, such as classical logic, fuzzy logic, multi-valued logic, Demster—Shafer theory, probabilistic logic, the logic of probability, Bayesian networks, semantic networks, rough sets, or default logic. Firstly, there are many ways of inference in LPR, which are not present in the formalisms mentioned above. Secondly, many parameters are specified for representing the uncertainty of knowledge. LPR is established on the basis of descriptions of reasoning carried out by people.

As the subtitle of the publication indicates ("A Core Theory"), during the development of formalism, attention is focussed on the most important elements of reasoning, which are identified in collected surveys. Minor issues such as reasoning related to time, space and meta-knowledge (knowledge about knowledge) have been left for further studies.

On the basis of LPR, a Dynamically Interlaced Hierarchies (DIH) formalism was developed (Hieb and Michalski 1993). Knowledge consists of a static part represented by hierarchies and a dynamic part, which are traces, playing a role similar to statements in LPR. The DIH distinguishes three types of hierarchies: types, components and priorities. The latter type of hierarchy can be divided into subclasses: hierarchies of measures (used to represent the physical quantities), hierarchies of quantification (allowing quantifiers to be included in traces, such as e.g. one, most, or all) and hierarchies of schemes (used as a means for the definition of multi-argument relationships and needed to interpret the traces).

Michalski et al. also developed a DIH implementation—an INTERLACE system. This programme can generate sequences of knowledge operations that will enable the derivation of a target trace from the input hierarchies and traces. Yet, not all kinds of hierarchy, probability and factors describing the uncertainty of the information were included there.

On the other hand, the studies done by Virvou et al. go in a different direction. LPR is used as a computer user modelling tool for user of computer systems and as a tool operating in intelligent tutoring systems. Studies described in Virvou and Du Boulay (1999) present RESCUER, a UNIX shell support system. By tracing changes in the file system and knowledge of the interpreter commands, the system is able to recognise the wrong commands and suggest appropriate substitutes. In Virvou (2002) LPR-based tutoring tool is presented.

Another field of formalism utilization is presented in the work done by Cawsey (1991). It discloses a system generating a description of the concepts based on the recipient's model, taking into account his/her current knowledge.

Research on LPR applications has been also performed at the AGH University. It concerned, in particular, diagnostics, knowledge representation and machine learning (Kluska-Nawarecka et al. 2014; Śnieżyński 2002; Górny et al. 2010).

3 Logic of Plausible Reasoning

As mentioned earlier, the LPR has been developed as a tool for modelling of human reasoning, and therefore some issues important for applications in the automated reasoning systems (e.g. variables) have not been taken into consideration. On the other hand, in the LPR, some solutions have been introduced that do not seem necessary and make the system implementation and creation of knowledge bases much more difficult. Below modifications are shown which adjust the LPR to the selected area of application. The modified formalism has been designated as LPR^0. This modification has been positively verified in practical application related to the construction of diagnostic systems and retrieval of information (Kluska-Nawarecka et al. 2014). The change most striking is the replacement of variable-valued logic, supplemented by parameters describing the uncertainty of information, with a labelled deductive system (LDS) (Gabbay 1991).

It is important to note that the deductive inference rules can correspond to induction (generalisation) at the level of knowledge of the world. This change is of an isomorphous character and allows for drawing of notation close to implementation (although knowledge is perhaps a little less human-readable). In addition, the change enabled a comprehensive, consistent and coherent approach to formalism: the knowledge base elements described by the parameters became the labelled formulas; transformations of knowledge, together with the methods of calculating the parameters of conclusions, have been superseded by the rules of inference, and the process of inference has been based on the construction of the evidences of formulas. With this approach, the apparatus of mathematical logic can be applied. Additionally to changing the notation, parts of formalism, which were considered unnecessary in this particular application, have been omitted as well as the related transformations. Some transformations were modified, also elements were added, such as object ordering, and the ability to use variables in formulas.

3.1 Language

The language used by LPR^0 consists of a countable set of constants C, variables X, the seven relational symbols, and logical connectives $\rightarrow$ and $\wedge$. It is formalized below:

Definition 1 The *language* of LPR^0 is a quadruple: $\mathrm{L} = (\mathrm{C}, \mathrm{X}, \{\mathrm{V}, \mathrm{H}, \mathrm{B}, \mathrm{E}, \mathrm{S}, \mathrm{P}, \mathrm{N}\}, \{\rightarrow, \wedge\})$

The set of constants C is used for representing concepts (objects) in the knowledge base. In certain cases, instead of fixed symbols, the symbols of variables can also be used, which enriches the LPR^0 language. The relational symbols (V, H, B, E, S, P, N) are used for defining relationships between concepts:

- H—three-argument relation defines the hierarchy between concepts; expression $\mathrm{H}(o_1, o, c)$ means that o_1 is o in a context c. Context is used for specification of the range of inheritance, o_1 and o have the same value for all attributes which depend on attribute c of object o;
- B—two-argument relation is used to present the fact that one object is placed below another one in a hierarchy, which is denoted as: $\mathrm{B}(o_1, o_2)$;
- V—three-argument relation is used for representing statements in the form of object-attribute—value relations; the notation in the form V o, a, v) is a

representation of the fact that object o has an attribute a equal to v. If object o has several values of a, several appropriate statements should be made in a knowledge base;

- E—four-argument relation is used for representing relationships; the notation E(o_1, a_1, o_2, a_2) means that values of attribute a_1 of object o_1 depend on attribute a_2 of the second object. To represent bidirectional dependency [mutual dependency, see Collins and Michalski (1989)], a pair of such expressions is needed;
- S—three-argument relation determines similarity between objects; S(o_1, o_2, c) represents the fact that o_1 is similar to o_2. Context, as above, specifies the range of similarity. Only these attributes of o_1 and o_2 which depend on attribute c will have the same value;
- P—two-argument relation represents order between concepts; P(o_1, o_2) means that concept o_1 precedes concept o_2;
- N—two-argument relation is used for comparing the concepts; N(o_1, o_2) means that concept o_1 is different from the concept o_2.

It should be noted that three new relations are introduced, namely B, N and P. The first of these relations is added to maintain the accuracy of the knowledge base after the use of generalisation or similarity of values (cf. Sect. 3.2), and it will not appear in the domain knowledge base. The second one may appear in the knowledge base as a premise in formulas having the form of implications. The last one, appearing in the knowledge base by itself, serves as a tool to represent order between the values and is used in implications to compare ordered values, which is often needed in practical applications. Using symbols of the relationship, one can define formulas first, and labelled formulas next, the latter ones being the basic elements of knowledge.

Definition 2 The *atomic formula* LPR^0 means every expression of the form of H(o_1, o_2, c), B(o_1, o_2), V(o, a, v), E(o_1, a_1, o_2, a_2), S(o_1, o_2, c), P(o_1, o_2), N(o_1, o_2), where $o, o_1, o_2, a, a_1, a_2, c, v \in C \cup X$

Elements of knowledge can also be formulas in the form of implications. To build expressions of this type, logical connectives are used. LPR uses a simplified form of the implications, called Horn clauses. They occur in the form of $\alpha_1 \wedge \alpha_2 \wedge \cdots \alpha_n \rightarrow \alpha$, $n \in$ N. It is also assumed that the components of implications can only be atomic formulas of the V, P and N type. Now the LPR formula can be defined.

Definition 3 *LPR formula* means every atomic formula, a conjunction of atomic formulas and an expression in the form of $\alpha_1 \wedge \alpha_2 \wedge \cdots \alpha_n \rightarrow$ V(o, a, v), where $n \in$ N, $n > 0$, α_i has the form of $V(o_i, a_i, v_i)$, $P(v_i, w_i)$ or $N(v_i, w_i)$, and o, o_i, a, a_i, v, v_i, $w_i \in C \cup X$ for $1 \le i \le n$. The set of all formulas is denoted by F.

To better illustrate the language, Table 1 shows examples of the atomic formulas related to the diagnosis of defects in castings (Kluska-Nawarecka 1999). The following are examples of formulas in the form of implications. From the original notation proposed by Collins and Michalski they differ in the method of

Table 1 Examples of atomic formulas

LPR^0 formula	Original form (LPR)
H(cast steel, material, properties)	Caststeel SPEC material in CX (material, properties (material))
S(cast steel, cast iron, mouldmaking)	Caststeel SIM castiron in CX (caststeel, mouldmaking (caststeel))
E(material, maxThicknessProtectiveCoating, material, mouldmaking)	maxThicknessProtectiveCoating (material) ↔ mouldmaking (material)
V(caststeel, minPermeability, permeabilityMedium)	minPermeability (caststeel) = permeabilityMedium
P(permeabilityMedium, permeabiltyHigh)	N/A

recording premises and conclusions, and in the fact that they may contain variables (capitalised):

$$\begin{aligned} &\mathrm{V(o, defect,\ pittedSkin)} \wedge \mathrm{V(protectiveCoating, thickness,}\ T) \\ &\quad \wedge \mathrm{P(thicknessCorrect,}\ T) \rightarrow \mathrm{V(defect, cause, tooLargeCoatingThickness)} \end{aligned} \tag{1}$$

$$\begin{aligned} &\mathrm{V(equipment, block,}\ B) \wedge \mathrm{V(}B\mathrm{, correctSignal,}\ P) \\ &\quad \wedge \mathrm{V(}B\mathrm{, currentSignal,}\ A) \wedge \mathrm{N(}P, A) \rightarrow \mathrm{V(damage, place,}\ B) \end{aligned} \tag{2}$$

During inference, the variables can be matched with different constants, owing to which the rules become more general.

In the first implication, the variable T represents actual thickness of the mould protective coating. This formula therefore represents the rule, which says that in the event of the occurrence of a defect called pitted skin, when the actual coating thickness exceeds the maximum limit, it can be the cause of this defect.

In the second implication (formula 6), the variable B is any block device, P represents a valid signal waveform for this block, while A shows the waveform actually obtained. This implication represents the rule, which says that if in a certain block the current waveform differs from the correct course, the block is damaged.

To take into account the parameters describing the uncertainty and incompleteness of knowledge, the presented formalism should be extended by adding the algebra of labels (Gabbay 1991).

Defnition 4 *Label algebra* LPR^0 denotes a pair $\mathcal{A} = (A, \{f_{ri}\})$, where A is a set of labels, while $\{f_{ri}\}$ is a set of functions defined on labels: $f_{r_i} : A^{|pr(r_i)|} \rightarrow A$, where $pr(r_i)$ are the premises of rule r_i (rules are described under Sect. 3.2).

Defnition 5 *Labeled formula* LPR^0 denotes a pair f: p, where f is formula, and p is label. The set of labeled formulas is denoted by F_E. Knowledge base is any finite set of labeled, formulas, not containing besides the implication premises N-type nor B-type formulas.

3.2 Inference Rules

In this section, the rules of inference are introduced; by using them one can construct the proof of formulas—the operation which corresponds to reasoning Each rule r_i has the following form:

$$\frac{\begin{array}{c}\alpha_1 : p_1 \\ \alpha_2 : p_2 \\ \vdots \\ \alpha_n : p_n\end{array}}{\alpha : p} \tag{3}$$

Labeled formulas α_i:p_i are called the premises of r_i rule, and formula α: p is called its conclusion. As mentioned earlier, with labels of the premises, one can calculate the label of conclusion, using label algebra $\mathcal{A}$:

$$p = f_{r_i}(p_1, p_2, \ldots, p_n) \tag{4}$$

Rules have been given short, symbolic names, allowing them to be quickly identified. The rules that are associated with the generalisation and abstraction are denoted by the symbol GEN, the rules associated with specialisation and concretisation of knowledge are denoted by SPEC. The rules which use the similarity between concepts are denoted by SIM, those using the law of transitivity are denoted by TRAN, whereas the *modus ponens* rule uses the symbol MP. Due to the fact that similar transformations of knowledge may relate to different types of formulas, indexes have been introduced explaining meaning of the operation performed. Relative to this criterion, the rules can be divided into 6 groups. The membership in a given group is indicated by the form of the last premise and conclusion of the rule (the exception is the H_B rule, whose only premise is an H-type relation).

Due to lack of space, only rules that operate on the statement will be discussed. Index attached to the name of the rule tells us what is transformed: o is an object, and v is the value. These rules are shown in Table 2.

Having introduced the rule, one can define the proof of the labelled formula φ from a set of labelled formulas K and the notion of a syntactic consequence.

Definition 6 By the *proof* of labelled formula φ from the set of formulas K we mean the tree P such that its root is φ, and for each vertex ψ:

- if ψ is a leaf, then $\psi \in K$ or ψ is an instance of a formula belonging to K;
- otherwise, there is a rule of inference such that the vertices ψ_1, …, ψ_k being descendants of ψ are premises r_i, while ψ is its consequence; label of ψ is calculated from the labels of r_i premises by means of f_{ri}.

A set of proofs is denoted by Π

Definition 7 We say that a labelled formula φ is *a syntactic consequence* of the set of labelled formulas K (denoted by $K \vdash \varphi$) if there is a proof in Π for φ from K.

Table 2 Rules operating on the statements

GEN_o	$\frac{\begin{array}{l}\mathrm{H}(o_1,o,c)\\ \mathrm{E}(o,a,o,c)\\ \mathrm{V}(o_1,a,v)\end{array}}{\mathrm{V}(o,a,v)}$	$SPEC_o$	$\frac{\begin{array}{l}\mathrm{H}(o_1,o,c)\\ \mathrm{E}(o,a,o,c)\\ \mathrm{V}(o,a,v)\end{array}}{\mathrm{V}(o_1,a,v)}$	SIM_o	$\frac{\begin{array}{l}\mathrm{S}(o_1,o_2,c)\\ \mathrm{E}(o_1,a,o_1,c)\\ \mathrm{V}(o_2,a,v)\end{array}}{\mathrm{V}(o_1,a,v)}$
GEN_v	$\frac{\begin{array}{l}\mathrm{H}(v_1,v,c)\\ \mathrm{E}(a,o,a,c)\\ \mathrm{H}(o_1,o,c_2)\\ \mathrm{B}(v,a)\\ \mathrm{V}(o_1,a,v_1)\end{array}}{\mathrm{V}(o_1,a,v)}$	$SPEC_v$	$\frac{\begin{array}{l}\mathrm{H}(v_1,v,c)\\ \mathrm{E}(a,o,a,c)\\ \mathrm{H}(o_1,o,c_1)\\ \mathrm{V}(o_1,a,v)\end{array}}{\mathrm{V}(o_1,a,v_1)}$	SIM_v	$\frac{\begin{array}{l}\mathrm{S}(v_1,v_2,c)\\ \mathrm{E}(a,o,a,c)\\ \mathrm{H}(o_1,o,c_2)\\ \mathrm{B}(v_1,a)\\ \mathrm{V}(o_1,a,v_2)\end{array}}{\mathrm{V}(o_1,a,v_1)}$
		MP	$\frac{\begin{array}{l}\alpha_1 \wedge \cdots \wedge \alpha_n \rightarrow\\ \mathrm{V}(o,a,v)\\ \alpha_1\\ \vdots\\ \alpha_n\end{array}}{\mathrm{V}(o,a,v)}$		

The practical use of an LPR formalism has been made possible by the development of appropriate inference algorithm LPA (Śnieżyński 2003) based on an AUTOLOGIC system developed by Morgan (1985). In order to limit the scope of the search and generate optimal evidence, an A* algorithm has been used.

4 LPR-Based Information System

The goal of the system is to provide a tool for searching items matching a description given by the user using item descriptions and background knowledge. Architecture of the system is presented in Fig. 1. Users' preferences, items descriptions and background knowledge are stored in a Knowledge Base (KB). When user searches for the information, it provides a query, which is translated into LPR query formula and proof searching algorithm is executed. It returns items matching the query sorted by label values.

Query q has the following form:

$$q = \mathrm{B}(X, \text{ items}) \wedge \mathrm{V}(X, a_1, v_1) \wedge \cdots \wedge \mathrm{V}(X, a_m, v_m) \tag{5}$$

where X represents an item, which the user is looking for, a_1, v_1, ..., a_m, v_m are attributes and its values used by the user in the query. Inference engine searches for objects matching the query. They should be placed below items in a hierarchy and have attribute values matching ones given in the query.

To provide a flexible match it is assumed that the label l_q of q is calculated as an average of statement labels returned by the proof algorithm. Results are sorted using l_q labels. To limit the search space, inference depth can be limited.

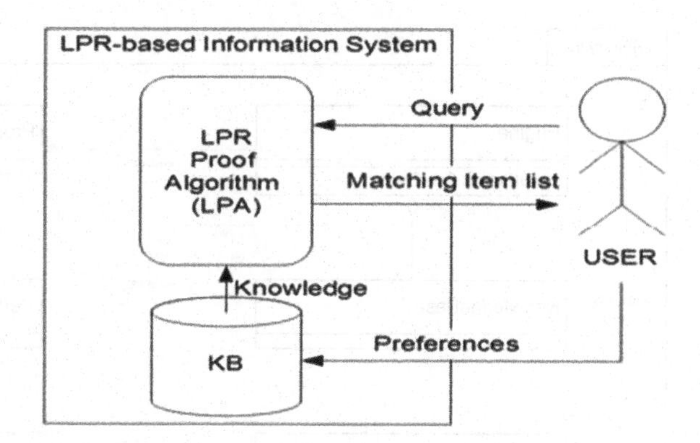

Fig. 1 Architecture of LPR-based information system

Experiments were executed with limit one (fast mode), which corresponds to the situation that item descriptions stored in KB should match the query, and with no limit, which allows to use domain knowledge (advanced mode).

User preferences are used to alter q label using knowledge about user's interests Knowledge base may contain information allowing to infer the statement of the form:

$$int = \mathrm{V}(U, \text{interested}, X) \tag{6}$$

where U represents the user, X represents item. Provided that $K \vdash int\text{: } l_{int}$ we can calculate resulting label as a sum: $l = l_q + l_{int}$.

User-related knowledge can have form of rules and take into account user's job (e.g. casting company), type of occupation (e.g. researcher or technologist), etc. They should be stored in the KB. Examples of rules are presented below:

$$\begin{aligned} &\mathrm{V}\,(U, \text{interested}, X) \leftarrow \mathrm{V}(X, \text{keyword, castingprocess}) \\ &\quad \wedge \mathrm{V}(U, \text{occupationtype, technologist}) \end{aligned} \tag{7}$$

$$\mathrm{V}(U, \text{interested}, X) \leftarrow \mathrm{V}(X, \text{material, adi}) \wedge \mathrm{V}(U, \text{job, castIndustry}) \tag{8}$$

The first rule can be interpreted as follows. The user is interested in item X if the article is about the casting process and the user is technologist. Second rule says that users working in the cast industry are interested in ADI material.

Software for the inference engine is developed in Java (Rawska et al. 2010). In Fig. 2 one can see the class diagram of the main package. `Engine` class is responsible for proof searching. `KnowledgeBase` stores all the knowledge available for the inference engine. `Proof` represents the proof tree being built. It is connected with subtrees by `Vertex` class. `Substitution` and `UnificationException` classes are used to process variables. To provide comfortable graphical user interface, web-based application was developed (Parada et al. 2012) using Google Web Toolkit (GWT). Its main query window is presented in Fig. 3. It is worth noting that the attributes can be single- or multi-valued. In the former case, the attribute has the nature of a unique description of the document, while in the latter case it can take several values from the given set.

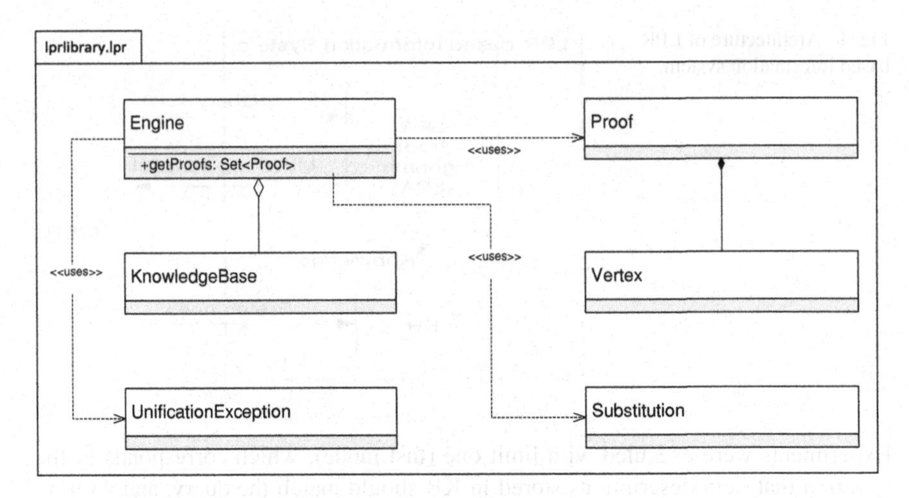

Fig. 2 Class diagram of the inference engine package

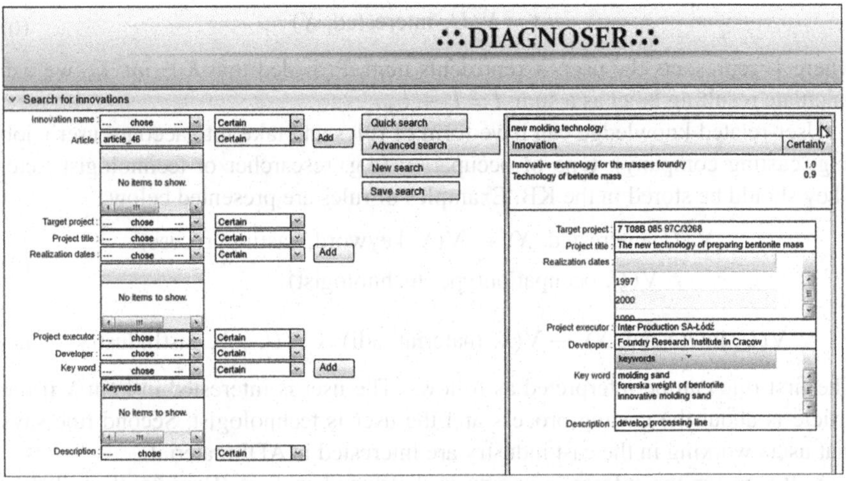

Fig. 3 Query window of the LPR-based intelligent information system (Parada et al. 2012)

In the process of searching, usually, only some of the attributes are known, while others are defined by the inference process conducted according to the LPR rules.

Creating a knowledge base consists of generating for each of the main entities a set of the corresponding formulas. This phase of building the knowledge base is performed by a knowledge engineer. It is outlined in the next section.

Table 3 Fragment of a list of innovations

No.	Name of innovation
1	Environment-friendly technology for the reclamation of bentonite moulding sands
2	New simplified method for the mechanical reclamation of moulding and core sands
3	Environment-friendly technology for the recycling of waste core sands
4	Development and implementation in production of an innovative design of the FT-65 impulse-squeeze moulding machine
5	New control and measurement apparatus adapted to the EU standards used for the bending strength determination in sands for the cold box process
6	An upgraded technology for the continuous preparation of moulding sands in a two-pan mixer

Table 4 List of attributes that describe documents related to innovation

Name of attribute	Type	Description
Innovation name	Single-valued	Name of innovation; every innovation has got a name
Article	Multi-valued	Articles about innovations, each article can also indicate a file or a link
Target project	Single-valued	Reference number of a target project
Project title	Single-valued	Title of project
Term of project	Multi-valued	Years in which the project was implemented, recorded in "YYYY" format
Project contractor	Single-valued	Name of company carrying out the project
Implementer	Single-valued	Project implementer
Keyword	Multi-valued	Keywords related to innovation
Description	Single-valued	Description of innovation

5 The Search for Innovative Casting Technologies

The area of domain knowledge, which was used to test the implemented prototype intelligent information system, is a knowledge of innovative casting technologies. Creating a knowledge base described in terms of LPR^0 required the participation of experts—in this case technologists from the Foundry Research Institute.

5.1 *Formal Description of Innovation*

As the first component of the knowledge, a statement of innovations that may be of interest to the user of the system was formulated. A fragment of this statement is shown in Table 3.

Another component, defined also with the participation of domain experts, is a list of attributes describing the documents under the consideration; a fragment of this list is shown in Table 4.

Table 5 Selected hierarchies in the description of innovation

Rule	Typicality and dominance	Description
H(*article_no*, article, name_innovation)	$\tau = 0.8$ $\delta = 0.5$	Each article is a type of article in the context of the name of a specific innovation
H(*article_no*, article, title_project)	$\tau = 0.8$ $\delta = 0.5$	Each article is a type of article in the context of the project title of a specific innovation
H(*name_contractor*, project contractor, project title)	$\tau = 0.8$ $\delta = 0.5$	Name of the project contractor is a type of the innovation contractor in the context of the project title
H(*key_word*, keyword, article)	$\tau = 0.8$ $\delta = 0.6$	Each keyword associated with innovation is a type of the keyword in the context of the article
H(*key_word*, keyword, project contractor)	$\tau = 0.8$ $\delta = 0.5$	Each keyword associated with innovation is a type of the keyword in the context of the project contractor
H(*key_word*, keyword, project title)	$\tau = 0.8$ $\delta = 0.5$	Each keyword associated with innovation is a type of the keyword in the context of the project title

Table 5 shows a fragment of the hierarchy statement established for the case of innovative technologies. In this table, the symbol H denotes hierarchical relationships, while words written in italics define variables for which values of the appropriate attributes are substituted.

A fragment of knowledge base regarding the similarity of concepts under consideration is presented in Table 6.

As before, italics in this table refer to the variables for which the values of attributes have been substituted.

A substantial part of the contents of the knowledge base is made of implications used in the process of reasoning. Selected implications describing the considered set of innovations are presented in Table 7. I is used instead of E to emphasise that entities represent innovations.

5.2 Selected Results of Tests

After entering the knowledge base of innovative foundry technologies into the information system, a number of experiments were carried out, and their aim was to test the functionality of the search.

The first group of tests consisted in searching for innovations defined by the user through different number of attributes. The search was conducted in both fast

Table 6 Selected formulas of similarities

Rule	Degree of similarity	Description
S(*key_word_1*, *key_word_2*, innovation)	$\sigma = 0.8$	Keyword 1 is similar to keyword 2 in the context of the same innovation
S(*key_word_1*, *key_word_2*, project title)	$\sigma = 0.8$	Keyword 1 is similar to keyword 2 in the context of the same project title
S(*key_word_1*, *key_word_2*, project contractor)	$\sigma = 0.8$	Keyword 1 is similar to keyword 2 in the context of the same project contractor
S(*article_1*, *article_2*, innovation)	$\sigma = 0.8$	Article 1 is similar to article 2 in the context of the same innovation
S(*article_1*, *article_2*, project contractor)	$\sigma = 0.8$	Article 1 is similar to article 2 in the context of the same project contractor of the same innovation
S(*name_contractor*, *name_implementer*, project title)	$\sigma = 0.8$	Contractor name is similar to the name of the implementer in the context of the project title of the same innovation

Table 7 Selected implications

Implication	Description
V(*U*, chose keyword, *Z*)∧ V(*I*, keyword, *Z*) → V(*U*, matches keyword, *I*):1.0	If user *U* chose keyword *Z* and innovation *I* has a link with keyword *Z*, then innovation *I* matches user *U* in the context of the keyword
V(*U*, chose completion date, *Y*)∧ V(*I*, completion date, Y) → V(*U*, matches completion date, *I*):1.0	If user *U* chose completion date *Y* and innovation *I* has a link with completion date *Y*, then innovation *I* matches user *U* in the context of the completion date
V(*U*, chose article, *M*)∧ V(*I*, article, *M*) → V(*U*, matches article, *I*):1.0	If user *U* chose article *M* and innovation *I* has a link with article *M*, then innovation *I* matches user U in the context of the article
V(*U*, chose project title, *N*)∧ V(*I*, project title, *N*) → V(*U*, matches project title, *I*):1.0	If user *U* chose project title *N* and innovation *I* has a link with project title *N*, then innovation *I* matches user U in the context of the project title
V(*U*, chose innovation name, *J*)∧ V(*I*, innovation name, *J*) → V(*U*, matches innovation name, *I*):1.0	If user *U* chose innovation name *J* and innovation *I* has a link with innovation name J, then innovation *I* matches user U in the context of the innovation name

(depth of the proof searching was limited to one) and advanced modes (depth was not limited). Selected results of these experiments are presented in Table 8 for one attribute chosen by the user and in Table 9 for the three attributes.

Table 8 Searching for innovations based on a single-attribute value

Attribute	Search mode	Searched innovations	Certainty of result
Attribute name: description; value: development and implementation in production of an innovative design of the FT-65 impulse-squeeze moulding machine; certainty: 0.75	Fast	Development and implementation in production of an innovative design of the FT-65 impulse-squeeze moulding machine to make moulds in bentonite sands	0.6
	Advanced	Development and implementation in production of an innovative design of the FT-65 impulse-squeeze moulding machine to make moulds in bentonite sands	0.499
		Implementation of the reclamation process of self-setting sands according to a new vibration-fluidised bed method	0.224
		Development and implementation of a technology for the manufacture of a new grade of foundry resin with reduced content of free formaldehyde and its application in selected domestic foundries	0.224
		Environment-friendly technology for the reclamation of bentonite moulding sands	0.224
Attribute name: article; value: article_5; certainty: 1.0	Fast	New simplified method for mechanical reclamation of moulding and core sands	1.0
		Environment-friendly technology for the reclamation of bentonite moulding sands	1.0
	Advanced	New simplified method for mechanical reclamation of moulding and core sands	1.0
		Environment-friendly technology for the reclamation of bentonite moulding sands	1.0
Attribute name: innovation name; value: environment-friendly technology for the reclamation of bentonite moulding sands; certainty: 0.75	Fast	Environment-friendly technology for the reclamation of bentonite moulding sands	1.0
	Advanced	Environment-friendly technology for the reclamation of bentonite moulding sands	0.75
		Implementation of the reclamation of self-setting sands according to a new vibration—fluidised bed method	0.375
		Development and implementation in production of an innovative design of the FT-65 impulse-squeeze moulding machine	0.375

Table 9 Searching for innovations based in plausible algebra calculations on the values of three attributes

Attributes	Type of search	Innovations search	Certainty of result
Attribute name: innovation name; value: new technology for casting cylinder sleeves in spinning metal moulds; certainty: 1.0	Fast	New technology for casting cylinder sleeves in spinning metal moulds	0.4666
		New simplified method for mechanical reclamation of moulding and core sands	0.4666
		Development and implementation in production of an innovative design of the FT-65 impulse-squeeze moulding machine to make moulds in bentonite sands	0.3333
Attribute name: project contractor; value: FERREX Sp. zo.o. Poznań; certainty: 1.0	Advanced	New technology for casting cylinder sleeves in spinning metal moulds	0.4666
		New simplified method for mechanical reclamation of moulding and core sands	0.4666
		Development and implementation in production of an innovative design of the FT-65 impulse-squeeze moulding machine	0.3333
		Development of design and implementation in production of a new versatile 20 dcm^3 capacity core shooter to make cores by various technologies	0.2999
Attribute name: project title; value: two-post FT-65 impulse-squeeze moulding machine; certainty: 1.0		Implementation of the reclamation process of self-setting sands according to a new vibration- fluidised bed method	0.2999
		Implementation of processes for moulding sand preparation, treatment and circulation and for casting automotive parts from ductile iron on a pilot line using prototype stand for impulse-squeeze moulding	0.2999

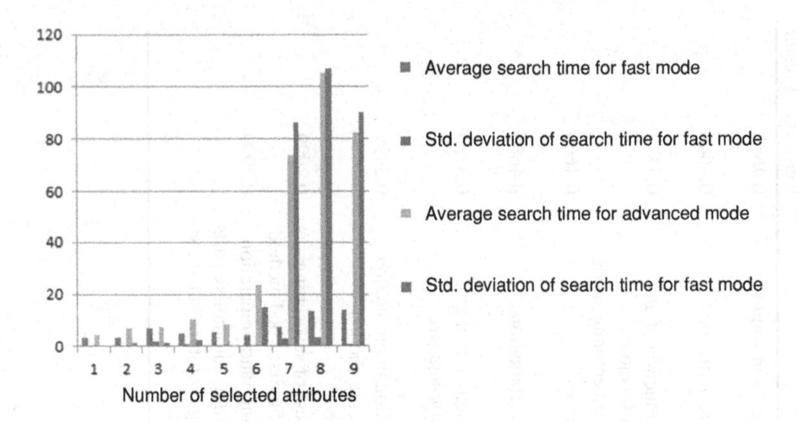

Fig. 4 Diagram of the time of waiting for the result depending on the number of selected attributes

Comparing the documents referred to in Table 8 (1 attribute) with the contents of Table 9 (3 attributes) it can be concluded that increasing the number of the included attribute values will generally result in their more accurate selection but does not always translate into increased certainty of the result.

Analysing the impact of the search mode, it is clear that the advanced mode usually leads to the selection of more documents than the fast mode (because reasoning is used to infer some of the attribute values), but the certainty of the result for the additionally indicated documents is usually lower (because of multiplication of values smaller than one).

In assessing the overall results of the tests, it can be concluded that they have demonstrated the correct operation of the tool constructed and allowed observing certain regularities found in the search for innovation.

The second part of the experiments concerned the performance parameters for which the following ones were adopted: the search time depending on the number of selected attributes and the search time depending on the number of specific innovations

Each search process was repeated three times for the same attribute values which were selected at random. Charts illustrating these relationships are presented in Figs. 4 and 5 and show the average values and standard deviations. As can be seen from the drawings, there are certain limits in the number of attributes and in the number of innovations beyond which the search time increases rapidly (especially when searching in advanced mode).

It is worth noting that, from a practical point of view, going in excess of these limits is not generally necessary, since it does not have a significant impact on a quality of the search result. However, if processing more information is necessary, a parallel implementation of the proof searching algorithm would be required.

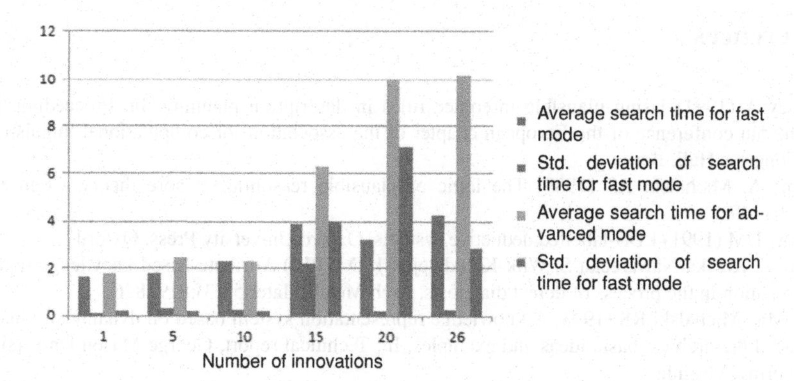

Fig. 5 Diagram of the time of waiting for the result depending on the size of the knowledge base

6 Conclusions

This chapter presents the concept of using LPR^0 to search for the text documents with a given theme. It seems that the proposed methodology for the description of documents greatly increases expression power in relation to commonly used search based only on keywords.

The attributes used in the described approach are a certain generalisation of keywords and, moreover, the use of LPR^0 formalism allows taking into account the context in which these attributes are present. Another advantage is possibility of using domain knowledge and inferring attribute values, which are not stored in the knowledge base explicit.

As a problem area for which the proposed methodology has been used, the search for innovation in casting technologies was chosen as characterising well enough the specific nature of this class of the tasks.

The designed pilot solution for the intelligent information tool makes it possible to carry out the experimental verification of the effectiveness of the approach. The tests performed are of course preliminary, but create successful outcomes for future work in this area.

In the future it is expected to examine the effectiveness of information retrieval with a much larger collection of source documents, and use data sources of a distributed character, the Internet included.

Very interesting also seems the possibility to equip the search engine with a learning procedure, which would allow for the automatic generation of a corresponding document in terms of LPR^0, consequently simplifying decisively the tedious process of creating a domain knowledge base.

Acknowledgments Scientific work financed from funds for the scientific research as an international project. Decision No. 820/N-Czechy/2010/0

References

Cawsey A (1991) Using plausible inference rules in description planning. In: Proceedings of the 5th conference of the European chapter of the association for computational linguistics, Congress Hall, Berlin

Collins A, Michalski RS (1989) The logic of plausible reasoning: a core theory. Cogn Sci 13:1–49

Gabbay DM (1991) LDS-labelled deductive systems. Oxford University Press, Oxford

Górny Z, Kluska-Nawarecka S, Wilk-Kolodziejczyk D (2010) Attribute-based knowledge representation in the process of defect diagnosis. Arch Metall Mater 55(3):819–826

Hieb MR, Michalski RS (1993) A knowledge representation system based on dynamically interlaced hierarchies: basic ideas and examples. In: Technical report, George Mason University, Fairfax, Virginia

Kluska-Nawarecka S (1999) Computer-aided methods for cast defects diagnosis. Foundry Research Institute, Krakow, Poland (in Polish)

Kluska-Nawarecka S, Śnieżyński B, Parada W, Lustofin M, Wilk-Kołodziejczyk D (2014) The use of LPR (logic of plausible reasoning) to obtain information on innovative casting technologies. Arch Civil Mech Eng 14(1):25–31

Michalski RS (1983) A theory and methodology of inductive learning. Artif Intell 20:111–161

Morgan CG (1985) Autologic. Logique et Analyse 28(110–111):257–282

Parada W, Lustofin M, Snieżyński B (supervisor) (2012) Expert system with learning capabilities based on the logic of plausible reasoning. In: Master Thesis, AGH University of Science and Technology, Poland (in Polish)

Rawska J, Rawski D, Snieżynski B (supervisor) (2010) Expert system with learning capabilities. In: Master Thesis, AGH University of Science and Technology, Poland (in Polish)

Śnieżyński B (2002) Probabilistic label algebra for the logic of plausible reasoning. In: Kłopotek M et al (eds) Intelligent information systems 2002 advances in soft computing. Springer, Berlin, pp 267–277

Śnieżyński B (2003) Proof searching algorithm for the logic of plausible reasoning. In: Kłopotek M et al (eds) Intelligent information processing and web mining advances in soft computing. Springer, Berlin, pp 393–398

Virvou M (2002) A cognitive theory in an authoring tool for intelligent tutoring systems. In: El Kamel A, Mellouli K, Borne P (eds), Proceedings of the IEEE international conference on systems, man and cybernetics

Virvou M, Du Boulay B (1999) Human plausible reasoning for intelligent help. User Model User-Adap Inter 9:321–375

Part II
Data Modeling, Acquisition and Mining

A Comparison of Two Approaches Used for Intelligent Planning of Complex Chemical Syntheses

Z. S. Hippe

Abstract Planning of complex chemical syntheses is illustrated by application of two different approaches (Corey's and Ugi-Dugundji's) in designing of new pharmaceuticals. The second approach (known worldwide as the **D–U** model) allows to generate synthetic pathways to required compounds, neglecting very expensive process of searching through enormously large databases on chemical reactions. The main features of both approaches are briefly discussed, and finally our own extension of the **D–U** model—based on the application of machine learning methods—is briefly discussed.

1 Introduction

In operational projects of the European Union related to high-level informational services for scientific, educational and industrial needs, planning syntheses of new medicines plays an important role. Nowadays, large intellectual and financial potential of international enterprises is applied along these lines. However, it seems that this potential does not keep track of the execution of recent challenges, because of the growing number of elderly people (requiring new, specific medicines), discovering of unknown illnesses, or existence of various bacteria mutating to forms resistant to regular antibiotics.

In the most recent way of planning new pharmaceuticals, all steps of the procedure are supported by advanced, sophisticated computer program systems. Some of these systems require an access to specialized databases (http://www.cas.org/, Accessed 5 Jan 2014) with information on properties of the investigated molecules, but above all—about methods of their syntheses and existing patent claims. However, despite of the fact that many researchers are deeply involved in creation

Z. S. Hippe (✉)
University of Information Technology and Management, Rzeszów, Poland
e-mail: zhippe@wsiz.rzeszow.pl

Z. S. Hippe et al. (eds.), *Issues and Challenges in Artificial Intelligence*,
Studies in Computational Intelligence 559, DOI: 10.1007/978-3-319-06883-1_7,

and exploitation of such databases, the process of their reliable search for required information is more and more difficult. This situation is caused by many factors; two of them are particularly frustrating. Firstly, at the present time we have huge number of analyzed objects: more than **~13 millions** of known substances and **several millions** of registered reactions. The second, very onerous factor, is the stereoisomerism of chemical molecules (Morrison and Boyd 1997).

Therefore, the following questions may be asked: *do we always have to search chemical databases during planning new syntheses? Can we directly generate required knowledge about chemical synthesis of a given compound?*

This chapter gives an attempt to reply to these questions. Thus we will talk about ***new*** interpretation of logic in design of chemical syntheses. This logic is additionally also new in the sense of approach to **H**uman-**S**ystem-**I**nteraction, while planning the syntheses of pharmaceuticals. Therefore, it seems necessary to get the acquaintance with a concept of *retrosynthetic* planning of reactions. *Retrosynthetic*—means reversely to the normal flow of a chemical reaction: {initial substance (substrate) → product}. Next, the use of **U**gi-**D**ugundji' *matrix model of constitutional chemistry* is briefly discussed. Finally, a unique reinforcement of the **D–U** model by selected methods of machine learning—developed in our research group—will be revealed.

2 Human-Controlled/Computer-Assisted Planning of New Pharmaceuticals

The traditional multi-step procedure used to obtain synthetically an organic substance, say a new medicine, is usually based on a sequence of the following operations:

(a) Discovery (frequently by a chance) an interesting medical activity of a given substance,
(b) Application of chromatographic methods to separate the active substance and to reject accompanying (ballasting) substances,
(c) Identification of a chemical structure of the active substance, usually using advanced spectral analysis,
(d) Searching through chemical and patent databases to get information about:
    - properties of homologous or similar structures and existing patent claims, *and*
    - known reactions used to synthesize the active substance,
(e) Proper planning of various synthetic pathways to synthesize the active substance (also, selecting the best route), *and finally*
(f) Realization the synthesis of a new medicine in laboratory scale, then in ¼- and ½-technical sizes, using Deming's stepwise method of quality optimization (Thompson and Koronacki 2001).

Sequential methodology described above does not enunciate clinical investigation of efficacy of the new medicine.

The key issue of the paper is related to the point **(e)** *planning of various synthetic pathways to synthesize the active substance*. The logic of **HSI**-methodology of planning complex strategies for chemical syntheses discussed here is based on entirely new ideas. Namely, two particular distinct methodologies to predict chemical syntheses of a compound having required structure (say, a medicine) can be applied, namely: (1) Corey's retrosynthetic computer-aided planning of chemical syntheses, and (2) Ugi-Dugundji' matrix model of constitutional chemistry. In both methodologies the searching of databases on chemical reactions can be completely neglected, at least in the step of planning syntheses. It gives us usually enormous benefits, for example—savings of time and money.

Detailed information about other items of the traditional procedure mentioned (points **(a)**, **(b)**, **(c)**, **(d)** and **(f)**) is available elsewhere (Bunin et al. 2007).

3 New Logic of Human-System-Interaction for Planning Chemical Syntheses

As it was stared earlier, there are two distinct logical approaches to predict chemical syntheses of a compound having required structure: (1) Corey's retrosynthetic computer-aided planning of chemical syntheses, *and* (2) Ugi–Dugundji' matrix model of constitutional chemistry.

- **Corey's Retrosynthetic Computer-Aided Planning of Chemical Syntheses**

In this approach to computer-aided planning of chemical syntheses, the molecule being synthesized (called the target molecule, **TM,** *Target Molecule*, a goal of the synthesis), is subjected to logical analysis in order to predict from which compounds it can be achieved in a single-step chemical conversions (i.e. reactions). These compounds form a set of *subgoals* of the first generation. Repetition of the procedure for each subgoal of the first generation results in the creation of subgoals of the second generation, etc. In this manner a tree-like structure, called a *synthesis tree*, is expanded. Particular *branches* and *nodes* in a synthesis tree represent *chemical conversions* (chemical reactions) and define *subgoals* (structures), respectively. The process of expanding a synthesis tree is continued until the generation of subgoal (subgoals) known to a chemist, or it is terminated automatically when the subgoal is readily available as a reactant for organic syntheses, laboratory or industrial. Generated in this manner *retrosynthetic pathways* (reversed paths of logical analysis) are simultaneously plans. They require further analysis in order to select plans with the most favorable characteristics (the smallest number of stages, each of them of the highest possible yield). This idea, developed by Nobel Prize winner EJ Corey from Harvard University, Cambridge, Massachusetts, USA, is applied in **LHASA** (http://www.infochem.de/. Accessed 5 Jan 2014), a system for computer-aided syntheses of carbogenes These substances are members of the family of carbon-containing compounds existing in an infinitive variety of compositions, forms and sizes. The naturally occurring carbogenes, or organic substances as they are known more traditionally, constitute the matter of all life on earth, and their science at the molecular level defines a fundamental language of that life. The idea of retrosynthesis was successfully applied to produce synthetic pathways for many important compounds. One spectacular synthesis having over 25 retrosynthetic steps was predicted for plant hormone giberrellic acid (its structural formula is not given here for obvious reasons). Particular importance of this compound, which increases crops of rice, is well understandable if we are thinking about starving Asia.

It is worth to emphasize that EJ Corey—a chemist, not only determined the working architecture of **LHASA**, but also made an important contribution to usage of trees as data-structures (Knuth 1997). Moreover, before it was announced by computer scientists, he proved that backward searching of trees is more effective than forward searching. However, one can say that **LHASA** has an ability of planning syntheses consisting solely of known (already described in the literature) reactions; only a sequence of these reactions (and/or model, **T**arget **M**olecule, for which they were applied) can be unlike.

- **Ugi-Dugundji' Matrix Model of Constitutional Chemistry**

The second logical approach, employing the matrix model of constitutional chemistry (further called the **D–U** model) (Dugundji and Ugi 1973; Hippe et al. 1992; Hippe 2011), is characterized by theoretical possibility to generate new, i.e. unprecedented chemical reactions. Hence, it appears that in the area of the simulation of chemical reactions, intelligent computer systems utilizing properly formulated phenomenal model and proper **HSI** user interface, are able to generate new, unprecedented knowledge. In no other field of application of artificial intelligence methods was it possible to achieve this level of solutions, using intelligent model based expert systems currently available. They usually are able only to *reproduce* knowledge contained in the knowledgebase(s) and unable to detect new facts or relations.

Coming back to questions issued in the Sect. 1, it seems necessary to recall about changes of philosophy of searching chemical databases for required information. The first breakthrough was done by German-Russian system **InfoChem** (http://nobelprize.org/nobel_prizes/chemistry/laureates/1990/corey-lecture.pdf, Accessed 5 Jan 2014). It offers a number of reaction searches (exact, substructure, similar, all-in-one) and a user-friendly editor to perform reaction queries easily. Moreover, instead standard searching through millions of chemical reactions, it is possible to search only for information about ***reaction type***. In the discussed system there are 41,300 *reaction types* altogether, one can say a decent number. Thus, instead of searching through more than millions of reactions, we can search for small number of reaction types. But a concept of the **D–U** model, as a much broader idea, allows to design chemical syntheses without searching of any database on chemical reactions. Here, a new data structure (called *reaction generators*) is employed for processing. Using standard *generate-and-test* algorithm (Jackson 1999), a real-time simulation of single-step chemical reactions is executed. The sequence of these reactions—meaningfully completed by an experienced chemist—can create a new, unprecedented synthetic pathways to the planned pharmaceutical. So, in this approach, Corey's retrosynthetic computer-aided planning of chemical syntheses is also performed, but in a completely different way.

Let us now give an outline of the theory of reaction generators. The most important for syntheses of carbogenes is the

2–2 generator (reads: two–two generator, i.e. *two* bonds *broken*, *two* bonds *made*).

$$\mathbf{A-B+C-D} \rightarrow \begin{cases} \mathbf{A-C+B-D} \\ \mathbf{A-D+B-C} \end{cases} \tag{1}$$

Objects **A**, **B**, **C** and **D**, shown in the notation (1), are *not* atoms, but some *fragments* of two molecules. Both bonds broken here (**A–B** and **C–D**) are active, automatically recognized within the structural formula of the analyzed molecule by algorithms for identification of substructures. The **2–2 generator** creates two different permutations. It was found, that they cover roughly **51 %** of currently known (described in chemical literature) organic reactions.

Originators of the matrix model described **42** kinds of reaction generators. Besides the **2–2 generator**, the following are very significant:

$$\mathbf{A{-}B + C{-}D + E{-}F \rightarrow A{-}C + D{-}E + B{-}F} \tag{2}$$

$$\mathbf{A{-}B + C :\rightarrow A{-}C{-}B} \tag{3}$$

These three reaction generators {(1), (2) and (3)} cover roughly **80 %** of useful chemical reaction applied in laboratory practice.

The key role in the matrix model of chemistry play matrices of *bonds-and-electrons*, **BE**, used for the description of reagents, i.e. objects participating in a chemical reaction. Another important feature blocks of the discussed model comprise *reaction matrices*, **R**. They represent classes of reactions (hence, indirectly—reaction generators). The mutual relations between matrices stated can be easily explained by an example of decomposition reaction of bromo-methanol to formaldehyde and hydro-bromide.

$$\mathbf{H^3{-}C^2(Br^6)(H^4){-}O^1{-}H^5 \rightarrow H^3{-}C^2(H^4){=}O^1 + H^5{-}Br^6} \tag{4}$$

	O^1	C^2	H^3	H^4	H^5	Br^6
O^1	4	1	0	0	1	0
C^2	1	0	1	1	0	1
H^3	0	1	0	0	0	0
H^4	0	1	0	0	0	0
H^5	1	0	0	0	0	0
Br^6	0	1	0	0	0	6

(B)

R →

O^1	C^2	H^3	H^4	H^5	Br^6
4	2	0	0	0	0
2	0	1	1	0	0
0	1	0	0	0	0
0	1	0	0	0	0
0	0	0	0	0	1
0	0	0	0	1	6

(E)

Bond-and-electron matrix of bromo-methanol (initial substance, educt), (**B**), (the character **B** stands for **B**egin) describes initial state of the reaction. On the other hand, *bond-and-electron* matrix, **E**, {from **E**nd}, shows that the reaction produces one molecule of formaldehyde and one molecule of hydro-bromide, both contained in one matrix.

So, to sum up, in the **D–U** model, structural formulas of the molecules constituting a chemical reaction are described in terms of bond-electron matrices, **BE**, and atomic vectors.

The matrix representation of chemical structures necessitates the assignment to each atom in a molecule a numerical identifier, the sequence of the numbering of atoms being arbitrary.

The **BE** matrix contains information on the way in which individual atoms are connected in a molecule and on the localization of free electrons (referring this information to identifiers), whereas the atomic vector conveys information on the kinds of atoms in a given chemical system (called also isomeric ensamble). Item (4) presents the atomic vector (**AV**, O^1 C^2 H^3 H^4 H^5 Br^6) and the **BE** matrices of molecules, taking part in the discussed reaction.

The bond-electron matrix has the following features (b_{ij} denotes a matrix element located in the ***i***-th row and ***j***-th column)**:** **(a)** it is a square matrix of dimensions $n \times n$, where n is the number of atoms present in molecules forming so called *ensamble of molecules*, **EM** (Dugundji and Ugi 1973); **(b)** each matrix element has a positive or zero value; **(c)** the numerical value of each off-diagonal element of the **BE** matrix (b_{ij}, $i \neq j$) represents a formal order of covalent bond between atoms A_i and A_j (numbers: **1**, **2** or **3** are allowed. When this value is equal to zero, the atoms A_i and A_j are not bound to each other); **(d)** values of diagonal elements (b_{ii}) stand for the number of free electrons of atom A.

More important features of the **D–U** model (say, detailed mathematical consequences) are given in Dugundji and Ugi (1973).

Provided that spatial arrangement of molecules is not considered, the atomic vector and BE matrix describe unambigously the structure of a chemical molecule. However, the disadvantage of the **BE** matrix are considerable redundancy of information, resulting from double notation of each bond, and large number of elements of a zero value (e.g. in the **BE** matrices in (1), zero entries constitute roughly **70 %** of all matrix elements). Therefore, it was introduced more concise method of describing the structure of chemical compound within the **D–U** model, involving the utilization of so-called list of atoms and list of bonds. The list of atoms, **LA**, describing the valence state of individual atoms of a molecule, comprise the atomic vector and the free electrons vector (this is a diagonal of the **BE** matrix). On the other hand, the list of bonds (**LB**), containing information about off-diagonal and non-zero elements of the **BE** matrix, describes the way the atoms are connected through covalent bonds.

Using elementary principles of matrix analysis, the difference between the final state of reaction (described by the bond-electron matrix of reaction product(s), $\mathbf{BE}_{(\mathbf{E})}$) and the initial state (bond-electron matrix of reactant(s), $\mathbf{BE}_{(\mathbf{B})}$, respectively) may be expressed by means of the next matrix of the model—the so-called reaction matrix.

Introducing notation:

$$\mathbf{BE}_{(\mathbf{E})} = E, \text{ and } \mathbf{BE}_{(\mathbf{B})} = B$$

we have:

$$\mathbf{E} - \mathbf{B} = \mathbf{R} \tag{5}$$

The reaction matrix ***R*** contains information about bonds which are broken and formed during the considered reaction, and about dislocations of free electrons in individual reagents. Rearrangement of the above equation to the form:

$$\mathbf{B} + \mathbf{R} = \mathbf{E} \tag{6}$$

yields the Ugi-Dugundji' *master equation*, representing notation of ***any*** organic reaction in the matrix form. The equation states that by adding the reaction matrix to the bond electron matrix of the reactant(s), the bond-electron matrix of the product(s) is obtained.

Let us now start with matrix ***B***, describing molecular structure of the analyzed molecule (say, molecule of the required drug), for which we try to get an expectation of possible synthetic pathways. In this case, the matrix ***B*** should be transformed by addition of **all** *reaction matrices* known, to get a family of all conceivable reactions:

$$\mathrm{E} \in \{\mathrm{E}_1, \mathrm{E}_2, \mathrm{E}_3, \ldots, \mathrm{E}_n\} \tag{7}$$

Association of the matrix ***B*** with molecules of so-called reaction partners (low molecular weight compounds, released or consumed during chemical reactions, e.g. H_2O, HCl, H_2, Cl_2, Br_2, O_2, CO_2, NH_3, CH_3OH, etc.) enables prediction of the conversions of a particular chemical system in various media (e.g. in water, in dry air, in moist soil, in a human body, etc.).

Numerous examples of practical applications of the **D–U** model allow us to draw a general conclusion that the known chemistry is but a small subset of conceivable chemistry. Chemical reactions—in other words, chemical knowledge—generated on the basis a mathematical model of constitutional chemistry may differ with respect to the degree of novelty: a reaction may be completely new in term of all aspects taken into consideration or it may be, belonging to a class of unknown conversions, a close analogue to some known chemical reactions.

4 Current Development of the Matrix Model

Continuous investigations are performed in the research group in Munich (Technische Universität Műnchen, **TUM**) and in Rzeszow (jointly at University of Information Technology and Management, and at Rzeszow University of Technology). Increasingly important research is done in Munich, devoted to application the **D–U** model in simulations of multi-component reactions. Other aspects of the **D–U** model are investigated in Rzeszow, mainly related to reaction matrices describing chemical conversions beyond the basic constitutional chemistry; i.e. with radicals, ions, electron gaps and electron pairs (**8** new kinds of reaction matrices are developed). But the most recent concept of human-system-interaction logic has been implemented in our *Chemical Sense **Builder***, **CSB,** running the **D–U** model. Frankly speaking, the selection of optimum strategy of chemical syntheses is rather troublesome, due to frequently divergent demands put on computer-assisted syntheses design systems, like **CSB**. Criteria for evaluation of strategy of synthesis of a given organic compound are different in the case of research & development (R&D) compared with those used in the prediction of industrial syntheses. Thus, the problem involves formulating general and universal principles, that would be convenient from the standpoint of the user of a such system, and can thus be adapted to discussion of a synthetic pathway, consistent with a natural course of synthesis under laboratory conditions. Apparently, at the outset, the optimum synthetic strategy may be defined as a sequence of chemical conversions which leads, with the highest possible yield, from readily available starting chemical (chemicals) to the desire compound. This definition is used here to expansion of a synthetic tree strictly in backward direction, i.e. according to Corey's philosophy. In this case, the term "sequence of chemical conversions" refers to selection of the best sequence of *logical transformations* of a target molecule and successive subgoals, leading to a readily available reactant or to a chemical molecule, using the methods of synthesis described in the literature. However, a fundamental difficulty, associated with the size of a state space and the possibility of its search, is encountered immediately. Namely, it is apparent that a correct choice of an optimum strategy can take place *only* as a result of *global* analysis of all existing solutions, which implies a necessity of prior generation of a *complete* synthetic tree. This is usually unattainable, due to an excessively large size of a solution space and/or limited speed of execution of logical and arithmetic operations by current computers (even super-computers). Hence, the optimum strategy of synthesis is necessarily determined on the basis of *local* analysis of the quality of suggested methods of creation of subgoals of a particular generation (level), not accounting for information about subgoals located at other levels of a synthetic tree. This signifies that selection of the first conversion is dictated *exclusively* by parameters of a target molecule, disregarding the effects implied by the structure of the subgoals of successive generations. Thus, it may happen that a formally promising strategy (with a respect to a target molecule) leads to a subgoal, for which a suitable transformation operator (chemical reaction) cannot be found: consequently, the synthetic pathway being advantageously developed in the initial stage of synthesis, becomes unexpectedly blocked. On the contrary, the strategy not appearing too promising at the first level of a synthetic tree, and thus subjected to instantaneous elimination, could have been—after considering subsequent stages—a quite acceptable starting point for initiation of an efficient synthetic pathway. To avoid these dangers, at least several plausible strategies are established for each node (i.e. for a target molecule and for all subgoals being successively generated), and their accomplishment *is executed*. In this relatively simple way, the *local* hierarchy of priorities of particular strategies can be determined with the simultaneous control of expansion of a synthetic tree. Its size usually correlates with the expected probability that among all generated synthetic pathways at least one will be promising, i.e. deserving laboratory validation. Local analysis of the quality of strategies selected by the **CSB** system (and, more strictly, evaluation of subgoals created in a given node due to application of these strategies) and elimination of less promising solutions result in the generation of the synthetic tree being considerably smaller compared with the complete tree; at the same time, it is anticipated that this restricted tree contains synthetic pathways, the majority of which can be considered valuable. However, the discussed methodology may by encumbered with a substantial risk of failure.

It was found, that the predictability of the **CSB** system can be distinctly increased using various algorithms of machine learning (Hippe 2007; Grzymala-Busse et al. 2010; Cohagan et al. 2012). As a matter of fact, machine learning is employed in order to reinforce the "second" part of the *generate-and-test* paradigm, i.e. for more sophisticated testing the chemical sense of generated reactions. Quoted here a collection of self-explanatory screen panels shows the exceptional predictability of **CSB** system, running the enhanced **D–U** model to predict reactions of Viagra (m1) with water (m2). In the first experiment all (50) reaction generators were applied; reaction enthalpy range (−100–30 Kcal/mole) was fixed; 382 reactions were predicted. Then, for the same set of reacting substances, a common sense reasoning model (it allows additionally selection of the reaction medium) was used, giving only 10 most reasonable chemical transforms.

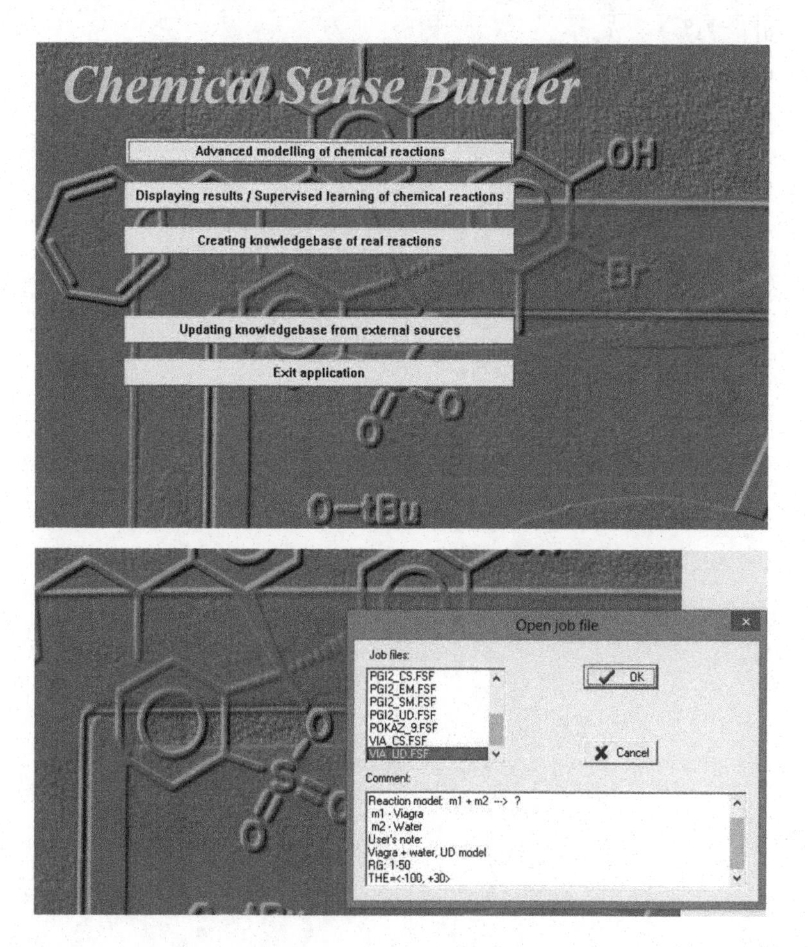

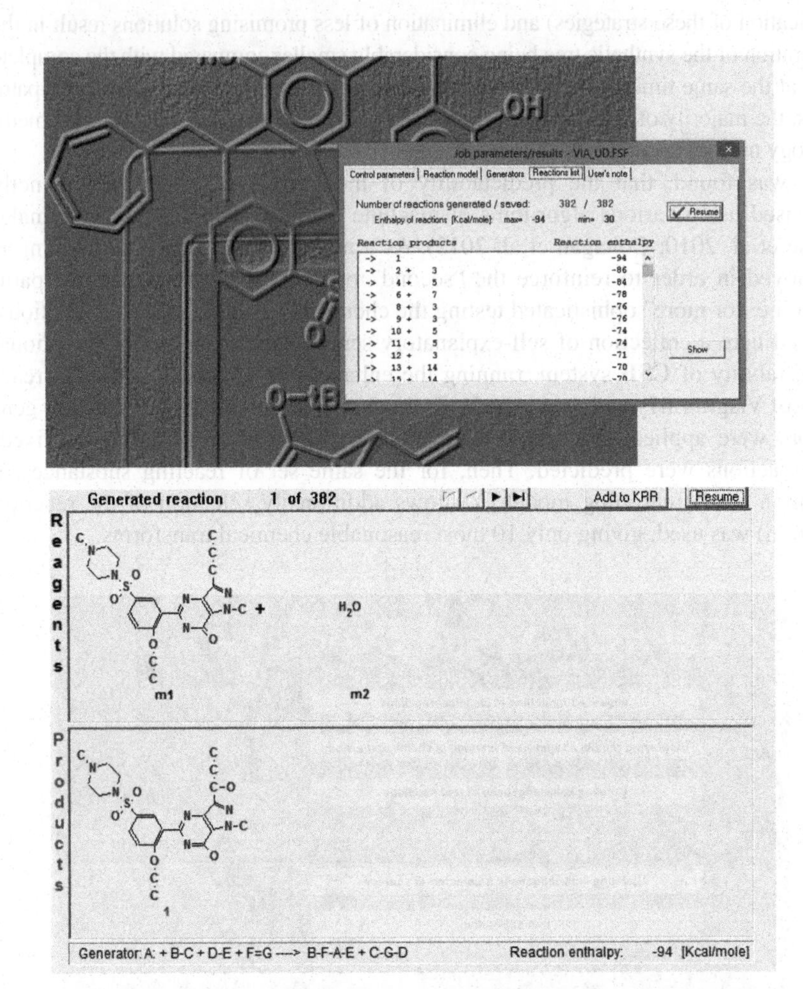
Job parameters/results - VIA_UD.FSF
Control parameters
Reaction model
Generators
Reactions list
User's note
Number of reactions generated / saved: 382 / 382
Enthalpy of reactions [Kcal/mole]: max= -94 min= 30
Resume
Reaction products
Reaction enthalpy
Show
Generated reaction 1 of 382
Add to KRR
Resume
Reagents
m1
m2
Products
Generator: A: + B-C + D-E + F=G ---> B-F-A-E + C-G-D
Reaction enthalpy: -94 [Kcal/mole]

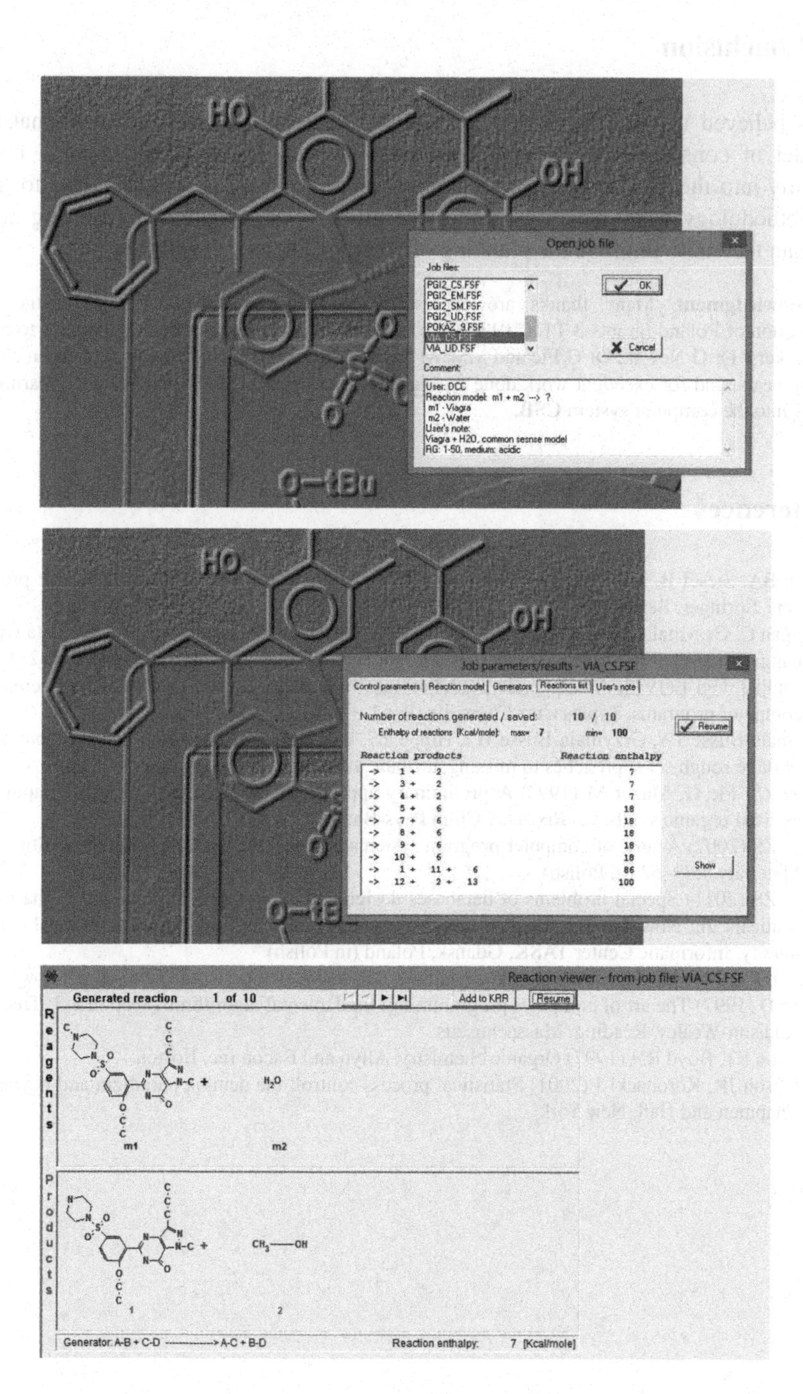

Open job file
Job files:
PGI2_CS.FSF
PGI2_EM.FSF
PGI2_SM.FSF
PGI2_UD.FSF
POKAZ_9.FSF
VIA_CS.FSF
VIA_UD.FSF
OK
Cancel
Comment:
User: DCC
Reaction model: m1 + m2 ---> ?
m1 - Viagra
m2 - Water
User's note:
Viagra + H2O, common sesnse model
RG: 1-50, medium: acidic
Job parameters/results - VIA_CS.FSF
Control parameters
Reaction model
Generators
Reactions list
User's note
Number of reactions generated / saved: 10 / 10
Enthalpy of reactions [Kcal/mole]: max= 7 min= 100
Resume
Reaction products
Reaction enthalpy
-> 1 + 2 7
-> 3 + 2 7
-> 4 + 2 7
-> 5 + 6 18
-> 7 + 6 18
-> 8 + 6 18
-> 9 + 6 18
-> 10 + 6 18
-> 1 + 11 + 6 86
-> 12 + 2 + 13 100
Show
Reaction viewer - from job file: VIA_CS.FSF
Generated reaction 1 of 10
Add to KRR
Resume
Reagents
m1
m2
Products
1
2
Generator: A-B + C-D ------> A-C + B-D
Reaction enthalpy: 7 [Kcal/mole]

5 Conclusion

It is believed that results of our experiments devoted to enhancement of matrix model of constitutional chemistry by machine learning algorithms, put a new quality into the **D–U** model. Its implementation in our **CSB** system seems to fill a methodology gap—which combines the efforts of knowledge engineering and human factors to improve the planning of complex chemical syntheses.

Acknowledgment Many thanks are expressed for generous support from Ministry of Education of Poland (grants 3 T11C 013 30 and 3 T11C 033 30). I am extremely grateful to my coworkers: Dr G Nowak, Dr G Fic and MSc M Mazur for creative and devoted cooperation over many years, and for excellent work done during implementation of some new machine learning ideas into the computer system **CSB.**

References

Bunin BA, Siesel B, Morales G, Bajorath J (2007) Chemoinformatics: theory, practice & products. Springer, Berlin

Cohagan C, Grzymala-Busse JW, Hippe ZS (2012) Experiments on mining inconsistent data with bagging and the MLEM2 rule induction. J Granular Comp, Rough Sets Int Syst 2:257–271

Dugundji J, Ugi I (1973) An algebraic model of constitutional chemistry as a basis for chemical computer programs. Topics Curr Chem 39:19–64

Grzymala-Busse JW, Grzymala-Busse WJ, Hippe ZS, Rząsa W (2010) An improved comparison of three rough set approaches to missing attribute values. Control Cybern 39:1–18

Hippe ZS, Fic G, Mazur M (1992) A preliminary appraisal of selected problems in computer-assisted organic synthesis. ReclTrav Chim Pays-Bas 111:255–261

Hippe ZS (2007) A suite of computer program systems for machine learning. Zamojskie Studia i Materiały 9:39–52 (in Polish)

Hippe ZS (2011) Special problems of databases design and exploitation in planning of pharmaceuticals. In: Nowakowski A (ed) Infobases: science, european projects, informational community. Informatic Center TASK, Gdańsk, Poland (in Polish)

Jackson P (1999) Introduction to expert systems. Addison-Wesley Longman Limited, Harlow

Knuth D (1997) The art of computer programming: fundamental algorithms (Section 2.3: Trees). Addison-Wesley, Reading, Massachusetts

Morrison RT, Boyd RN (1997) Organic chemistry. Allyn and Bacon Inc, Boston

Thompson JR, Koronacki J (2001) Statistical process control: the deming paradigm and beyond. Chapman and Hall, New York

Mining Inconsistent Data with Probabilistic Approximations

P. G. Clark, J. W. Grzymala-Busse and Z. S. Hippe

Abstract Generalized probabilistic approximations, defined using both rough set theory and probability theory, are studied using an approximation space (U, R), where R is an arbitrary binary relation. Generalized probabilistic approximations are applicable in mining inconsistent data (data with conflicting cases) and data with missing attribute values.

1 Introduction

Rough set theory is based on ideas of lower and upper approximations. Such approximations are especially useful for handling inconsistent data sets. Complete data sets, i.e., data in which every attribute value is specified, are described by an indiscernibility relation which is an equivalence relation. For complete data sets an idea of the approximation was generalized by introducing probabilistic approximations, with an additional parameter, interpreted as a probability (Yao 2007; Yao and Wong 1992).

P. G. Clark · J. W. Grzymala-Busse (✉)
Department of Electrical Engineering and Computer Science, University of Kansas, Lawrence, KS, USA
e-mail: jerzy@eecs.ku.edu

P. G. Clark
e-mail: patrick.g.clark@gmail.com

J. W. Grzymala-Busse
Institute of Computer Science, Polish Academy of Sciences, 01-237 Warszawa, Poland

Z. S. Hippe
Department of Expert Systems and Artificial Intelligence, University of Information Technology and Management, 35-225 Rzeszów, Poland
e-mail: zhippe@wsiz.rzeszow.pl

Z. S. Hippe et al. (eds.), *Issues and Challenges in Artificial Intelligence*,
Studies in Computational Intelligence 559, DOI: 10.1007/978-3-319-06883-1_8,

For incomplete data sets there exist three definitions of approximations, called singleton, subset and concept. Probabilistic approximations were extended to incomplete data sets by introducing singleton, subset and concept probabilistic approximations in Grzymala-Busse (2011). First results on experiments on probabilistic approximations were published in Clark and Grzymala-Busse (2011). Lately, an additional type of probabilistic approximations, called local, was introduced in Clark et al. (2012).

In incomplete data sets, there are two types of missing attribute values: lost values and "do not care" conditions (Grzymala-Busse 2003). Lost values mean that the original attribute values are not available, e.g., they were erased or the operator forgot to insert them into the data set. In the process of data mining we will use only existing attribute values. On the other hand, "do not care" conditions are interpreted as any possible existing values from the attribute domain, i.e., we are assuming that such a value may be replaced by any attribute value.

A preliminary version of this chapter was prepared for the HSI 2013, 6-th International Conference on Human System Interaction, Gdansk, Poland on June 8, 2013 (Grzymala-Busse 2013).

2 Complete Data Sets

Our basic assumption is that the data sets are presented in the form of a *decision table*. An example of a decision table is shown in Table 1. Rows of the decision table represent *cases*, while columns are labeled by *variables*. The set of all cases is denoted by U. In Table 1, $U = \{1, 2, 3, 4, 5, 6, 7, 8\}$. Some variables are called *attributes* while one selected variable is called a *decision* and is denoted by d. The set of all attributes will be denoted by A. In Table 1, $A = \{$*Wind, Humidity, Temperature*$\}$ and $d =$ *Trip*. For an attribute a and case x, $a(x)$ denotes the value of the attribute a for case x. For example, *Wind* (1) = *low*.

A significant idea used for scrutiny of data sets is a *block of an attribute-value pair*. Let (a, v) be an attribute-value pair. For *complete* data sets, i.e., data sets in which every attribute value is specified, a block of (a, v), denoted by $[(a, v)]$, is the following set.

$$\{x | a(x) = v\} \tag{1}$$

For Table 1, blocks of all attribute-value pairs are as follows:

[(Wind, low)] = {1, 2, 4}
[(Wind, high)] = {3, 5, 6, 7, 8}
[(Humidity, low)] = {1, 2, 3, 6, 8}
[(Humidity, high)] = {4, 5, 7}
[(Temperature, low)] = {2, 5, 7, 8}
[(Temperature, high)] = {1, 3, 4, 6}

A special block of a decision-value pair is called a *concept*. In Table 1, the concepts are [(Trip, yes)] = {1, 2, 3, 4, 5, 6} and [(Trip, no)] = {7, 8}.

Let B be a subset of the set A of all attributes. Complete data sets are characterized by the indiscernibility relation $IND(B)$ defined as follows: for any $x, y \in U$,

Table 1 A complete and inconsistent data set

Attributes				Decision
Case	Wind	Humidity	Temperature	Trip
1	Low	Low	High	Yes
2	Low	Low	Low	Yes
3	High	Low	High	Yes
4	Low	High	High	Yes
5	High	High	Low	Yes
6	High	Low	High	Yes
7	High	High	Low	No
8	High	Low	Low	No

$$(x, y) \in IND(B) \text{ if and only if } a(x) = a(y) \text{ for any } a \in B \tag{2}$$

Obviously, *IND*(*B*) is an equivalence relation. The equivalence class of *IND*(*B*) containing $x \in U$ will be denoted by $[x]_B$ and called *B*-elementary set. *A*-elementary sets will be called *elementary*. We have

$$[x]_B = \{[(a, a(x))] | \, a \in B\} \tag{3}$$

The set of all equivalence classes $[x]_B$, where $x \in U$, is a partition on *U* denoted by B^*. For Table 1, $A^* = \{\{1\}, \{2\}, \{3, 6\}, \{4\}, \{5, 7\}, \{8\}\}$. All members of A^* are elementary sets.

The data set, presented in Table 1, contains *conflicting cases*, for example cases 5 and 7: for any attribute $a \in A$, a(5) = a(7), yet cases 5 and 7 belong to two different concepts. A data set containing conflicting cases will be called *inconsistent*. We may recognize that a data set is inconsistent comparing the partition A^* with a partition of all concepts: there exists an elementary set that is not a subset of any concept. For Table 1, the elementary set {5, 7} is not a subset of any of the two concepts {1, 2, 3, 4, 5, 6} and {7, 8}.

The B-*lower approximation* of X, denoted by $\underline{appr}_B(X)$, is defined as follows

$$\{x | x \in U, [x]_B \subseteq X\} \tag{4}$$

The B-*upper approximation* of X, denoted by $\overline{appr}_B(X)$, is defined as follows

$$\{x \,| x \in U, [x]_B \cap X \neq \emptyset\} \tag{5}$$

For the data set from Table 1 and the concept [(Trip, yes)] = {1, 2, 3, 4, 5, 6} = X, $\underline{appr}_A(X) = \{1, 2, 3, 4, 6\}$ and $\overline{appr}_A(X) = \{1, 2, 3, 4, 5, 6, 7\}$.

2.1 Probablistic Approximations

Let (*U, R*) be an approximation space, where *R* is an equivalence relation on *U*. A probabilistic approximation of the set *X* with the threshold α, $0 < \alpha \leq 1$, is denoted by $appr_\alpha(X)$ and defined by

Table 2 Data sets used for experiments

Data set	Number of			Inconsistency
	Cases	Attributes	Concepts	Level (in %)
Australian	690	14	2	24.35
Echocardiogram	74	7	2	64.86
German	1,000	24	2	29.50
Glass	214	9	6	44.86
Hepatitis	155	19	2	34.19
Image segmentation	210	19	7	40.95
Iris	150	4	3	44.00
Postoperative patient	90	8	3	15.56
Primary tumor	339	17	21	27.73
Wine recognition	178	13	3	38.20

$$\cup\{[x] | x \in U, Pr(X \mid [x]) \geq \alpha\} \tag{6}$$

where [x] is an elementary set of R and Pr(X | [x]) is the conditional probability of X given [x].

$$Pr(X \mid [x]) = \frac{|X \cap [x]|}{|[x]|} \tag{7}$$

Obviously, for the set *X*, the probabilistic approximation of *X* computed for the threshold equal to the smallest positive conditional probability *Pr*(*X* | [*x*]) is equal to the standard upper approximation of *X*. Additionally, the probabilistic approximation of *X* computed for the threshold equal to 1 is equal to the standard lower approximation of *X*.

2.2 Experiments

For our experiments we used ten real-life data sets that are available on the University of California at Irvine *Machine Learning Repository*. These data sets were modified by discretization and by changing the inconsistency level, see Table 2. For rule induction we used two versions (global and local) of the MLEM2 (Modified Learning from Examples Module version 2) rule induction algorithm. Both versions were described in Grzymala-Busse and Rzasa (2010).

The main objective of our research was to test whether proper probabilistic approximations are better than lower and upper approximations. We conducted experiments of a single ten-fold cross validation starting with 0.001 and then increasing the parameter α by 0.1 until reaching 1.0. In our experiments, probabilistic approximations associated with α between 0.1 and 0.9 are called proper.

Results of our experiments are shown in Figs. 1, 2, 3, 4 and 5. In ten out of twenty cases proper probabilistic approximations were better than lower and upper

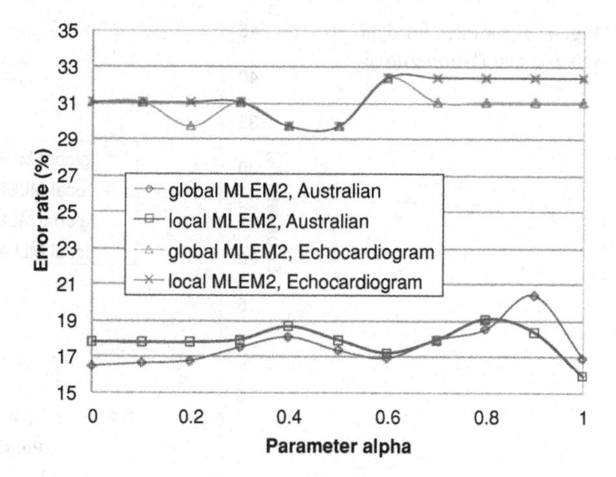

Fig. 1 Error rates for data sets *Australian* and *Echocardiogram*

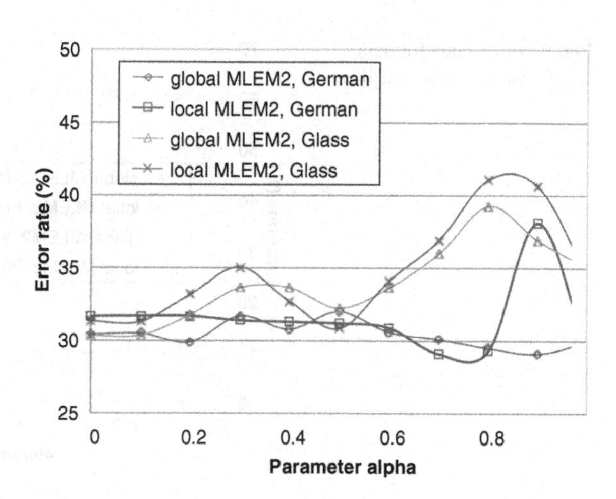

Fig. 2 Error rates for data sets *German* and *Glass*

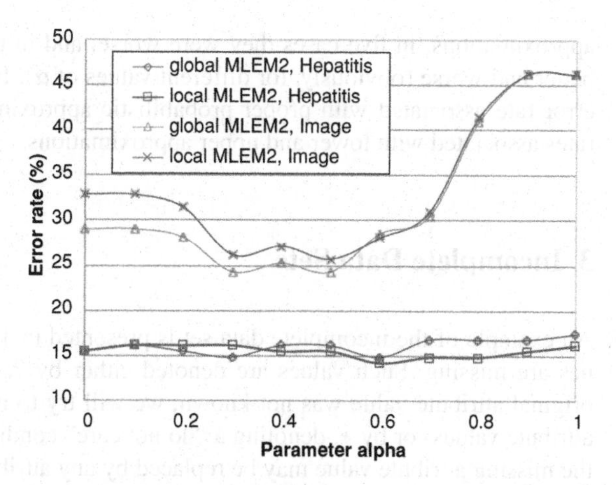

Fig. 3 Error rates for data sets *Hepatitis* and *Image*

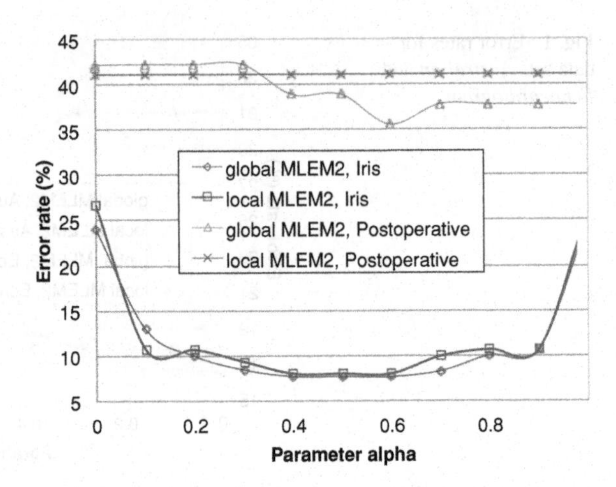

Fig. 4 Error rates for data sets *Iris* and *Postoperative*

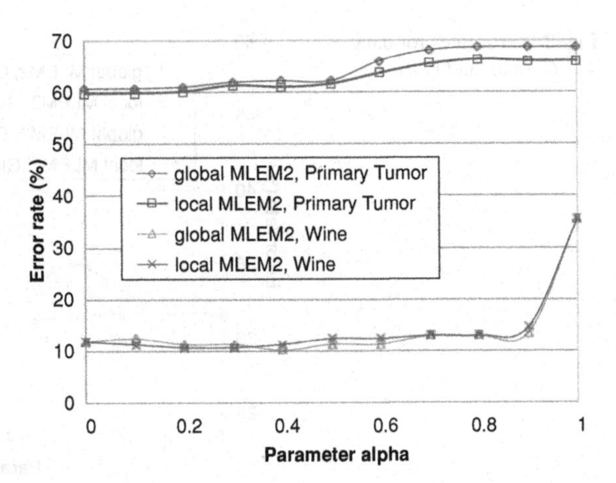

Fig. 5 Error rates for data sets *Primary Tumor* and *Wine*

approximations, in five cases they were worse, and in three cases they were both better and worse (obviously, for different values of α). In remaining two cases, the error rate associated with proper probabilistic approximations was between error rates associated with lower and upper approximations.

3 Incomplete Data Sets

An example of the incomplete data set is presented in Table 3. Some attribute values are missing. Such values are denoted either by ?, denoting a lost value (the original attribute value was not known, we will try to use only existing, specified attribute values) or by $*$, denoting a "do not care" condition (we are assuming that the missing attribute value may be replaced by any attribute value).

Table 3 An incomplete data set

Attributes				Decision
Case	Wind	Humidity	Temperature	Trip
1	Low	Low	High	Yes
2	?	Low	Low	Yes
3	High	?	High	Yes
4	Low	*	*	Yes
5	*	High	Low	Yes
6	High	Low	High	Yes
7	High	High	?	No
8	?	*	Low	No

For incomplete decision tables the definition of a block of an attribute-value pair is modified (Grzymala-Busse 2003, 2004).

- If for an attribute a there exists a case *x* such that *a*(*x*)= ?, i.e., the corresponding value is lost, then the case *x* should not be included in any blocks [(*a*, *v*)] for all values *v* of attribute *a*,
- If for an attribute *a* there exists a case *x* such that the corresponding value is a "do not care" condition, i.e., $a(x) = *$, then the case *x* should be included in blocks [(*a*, *v*)] for all specified values *v* of attribute *a*.

Thus, for Table 3, the blocks of all attribute-value pairs are as follows:

[(Wind, low)] = {1, 4, 5}
[(Wind, high)] = {3, 5, 6, 7}
[(Humidity, low)] = {1, 2, 4, 6, 8}
[(Humidity, high)] = {4, 5, 7, 8}
[(Temperature, low)] = {2, 4, 5, 8}
[(Temperature, high)] = {1, 3, 4, 6}

Let *B* be a subset of the set *A* of all attributes. For a case $x \in U$ the *characteristic set* $K_B(x)$ is defined as the intersection of the sets *K*(*x*, *a*), for all $a \in B$, where the set *K*(*x*, *a*) is defined in the following way:

If *a*(*x*) is specified, then *K*(*x*, *a*) is the block [(*a*, *a*(*x*))] of attribute a and its value *a*(*x*),

If $a(x) = ?$ or $a(x) = *$ then the set $K(x, a) = U$.

Characteristic set $K_B(x)$ may be interpreted as the set of cases that are indistinguishable from *x* using all attributes from *B*, with a given interpretation of missing attribute values. Thus, $K_A(x)$ is the set of all cases that cannot be distinguished from x using all attributes. For Table 3 and B = A, sets are as follows:

$K_A(1) = \{1, 4\}$
$K_A(2) = \{2, 4, 8\}$
$K_A(3) = \{3, 6\}$
$K_A(4) = \{1, 4, 5\}$
$K_A(5) = \{4, 5, 8\}$
$K_A(6) = \{6\}$
$K_A(7) = \{5, 7\}$
$K_A(8) = \{2, 4, 5, 8\}$

The characteristic sets K(B) uniquely define a characteristic relation R(B) on U defined for x,y ∈ U as follows $(x, y) \in R(B)$ if and only if $y \in K_B(x)$.

The characteristic relation *R*(*B*) is reflexive but—in general—does not need to be symmetric or transitive. In our example, *R*(*A*) = {(1, 1) (1, 4) (2, 2) (2, 4) (2, 8) (3, 3) (3, 6) (4, 1) (4, 4) (4, 5) (5, 4) (5, 5) (5, 8) (6, 6) (7, 5) (7, 7) (8, 2) (8, 4) (8, 5) (8, 8)}. A pair (*U*, *R*(*B*)) is also called an approximation space.

For incomplete data sets there exist three distinct definitions of approximations. Let X be a subset of U. The B-*singleton lower approximation* of X, denoted by $\underline{appr}_B^{singleton}(X)$, is defined as follows

$$\{x | x \in U, K_B(x) \subseteq X\} \quad (8)$$

The singleton lower approximations were studied in many papers, see, e.g. Grzymala-Busse (2003, 2004), Kryszkiewicz (1995), Slowinski and Vanderpooten (2000), Stefanowski and Tsoukias (1999), Yao (1998).

The B-*singleton upper approximation* of X, denoted by $\overline{appr}_B^{singleton}(X)$, is defined as follows

$$\{x | x \in U, K_B(x) \cap X \neq \emptyset\} \quad (9)$$

The singleton upper approximations, like singleton lower approximations, were also studied in many papers, e.g. Grzymala-Busse (2003, 2004), Kryszkiewicz (1995), Slowinski and Vanderpooten (2000), Stefanowski and Tsoukias (1999), Yao (1998).

The B-*subset lower approximation* of X, denoted by $\underline{appr}_B^{subset}(X)$, is defined as follows

$$\cup\{K_B(x) | x \in U, K_B(x) \subseteq X\} \quad (10)$$

The subset lower approximations were introduced in Grzymala-Busse (2003, 2004).

The B-*subset upper approximation* of X, denoted by $\overline{appr}_B^{subset}(X)$, is defined as follows

$$\cup\{K_B(x) | x \in U, K_B(x) \cap X \neq \emptyset\} \quad (11)$$

The subset upper approximations were introduced in (Grzymala-Busse 2003, 2004).

The B-*concept lower approximation* of X, denoted by $\underline{appr}_B^{concept}(X)$, is defined as follows

$$\cup\{K_B(x) | x \in X, K_B(x) \subseteq X\} \quad (12)$$

The concept lower approximations were introduced in Grzymala-Busse (2003, 2004).

The B-*concept upper approximation* of X, denoted by $\overline{appr}_B^{concept}(X)$, is defined as follows

$$\cup\{K_B(x) | x \in X, K_B(x) \cap X \neq \emptyset\} = \cup\{K_B(x) | x \in X\} \quad (13)$$

The concept upper approximations were studied in Grzymala-Busse (2003, 2004).

For Table 3 and X = {1, 2, 3, 4, 5, 6}, all A-singleton, A-subset and A-concept approximations are as follows:

$\underline{appr}_B^{singleton}(X) = \{1, 3, 4, 6\}$ $\qquad$ $\overline{appr}_B^{subset}(X) = U$
$\overline{appr}_B^{singleton}(X) = U$ $\qquad$ $\underline{appr}_B^{concept}(X) = \{1, 3, 4, 5, 6\}$
$\underline{appr}_B^{subset}(X) = \{1, 3, 4, 5, 6\}$ $\qquad$ $\overline{appr}_B^{concept}(X) = \{1, 2, 3, 4, 5, 6, 8\}$

4 Definability

A set X will be called *B-globally definable* if it is a union of some characteristic sets $K_B(x)$, $x \in U$. A set that is A-globally definable will be called *globally definable*.

A set T of attribute-value pairs, where all attributes belong to set B and are distinct, will be called a *B-complex*. Any A-complex will be called—for simplicity—a *complex*. A block of B-complex T, denoted by $[T]$, is defined as the set $\cap\{[t] \mid t \in T\}$. We are assuming that for any complex T the block of T is always not empty.

For an incomplete decision table and a subset B of A, a union of intersections of attribute-value pair blocks of attribute-value pairs from some B-complexes, will be called a *B-locally definable* set. A-*locally definable* sets will be called locally definable. For example, for the data set from Table 3, the set {4, 5} is locally definable, since {4, 5} = [(Wind, low)] ∩ [(Humidity, high)]) but it is not globally-definable. On the other hand, the set {8} is not even locally definable since all blocks of attribute-value pairs containing the case 8 contain the case 4 as well.

Any set X that is B-globally definable is B-locally definable, although the converse is not true.

5 Global Probabilistic Approximations

By analogy with lower and upper approximations defined using characteristic sets, we will introduce three kinds of probabilistic approximations: singleton, subset and concept. Again, let B be a subset of the attribute set A and X be a subset of U.

A B-singleton probabilistic approximation of X with the threshold with the threshold α, $0 < \alpha \leq 1$, is denoted by $appr_{\alpha,B}^{singleton}(X)$, is defined as follows

$$\{x|x \in U, Pr(X|K_B(x)) \geq \alpha\}, \tag{14}$$

$$Pr(X \mid K_B(x)) = \frac{|X \cap K_B(x)|}{|K_B(x)|}, \tag{15}$$

where Eq. 15 is the conditional probability of X given $K_B(x)$ and $|Y|$ denotes the cardinality of set Y.

A B-subset probabilistic approximation of X with the threshold with the threshold α, $0 < \alpha \leq 1$, is denoted by $appr_{\alpha,B}^{subset}(X)$, is defined as follows

$$\cup\{K_B(x)|x \in U, Pr(X|K_B(x)) \geq \alpha\} \tag{16}$$

A B-concept probabilistic approximation of X with the threshold with the threshold α, $0 < \alpha \leq 1$, is denoted by $appr_{\alpha,B}^{concept}(X)$, is defined as follows

$$\cup\{K_B(x)|x \in X, Pr(X|K_B(x)) \geq \alpha\} \tag{17}$$

Let *type* ∈ {*singleton, subset, concept*}. Note that

$$appr^{type}_{1,B}(X) = \underline{appr}^{type}_{B}(X) \tag{18}$$

and for the smallest possible positive α (in our experiments such α was equal to 0.001)

$$appr^{type}_{\alpha,B}(X) = \overline{appr}^{type}_{B}(X) \tag{19}$$

For Table 3, all distinct A-singleton, A-subset and A-concept approximations of the set $X = \{1, 2, 3, 4, 5, 6\}$ are as follows:

$appr^{singleton}_{0.5,A}(X) = U$

$appr^{singleton}_{0.667,A}(X) = \{1, 2, 3, 4, 5, 6, 8\}$

$appr^{singleton}_{0.75,A}(X) = \{1, 3, 4, 6, 8\}$

$appr^{singleton}_{1,A}(X) = \{1, 3, 4, 6\}$

$appr^{subset}_{0.5,A}(X) = U$

$appr^{subset}_{0.667,A}(X) = appr^{subset}_{0.75}$
$(X) = \{1, 2, 3, 4, 5, 6, 8\}$

$appr^{subset}_{1,A}(X) = \{1, 3, 4, 5, 6\}$

$appr^{concept}_{0.5,A}(X) = appr^{concept}_{0.667,A}$
$(X) = \{1, 2, 3, 4, 5, 6, 8\}$

$appr^{concept}_{0.75,A}(X) = appr^{concept}_{1,A}$
$(X) = \{1, 3, 4, 5, 6\}$

Notably, $appr^{singleton}_{0.75,A}(\{1, 2, 3, 4, 5, 6\}) = \{1, 3, 4, 6, 8\}$ is not even locally definable because the smallest intersections of attribute-value blocks that contain the case 8 must also contain {2, 4, 8} or {4, 5, 8}.

6 Local Probabilistic Approximations

Singleton, subset and concept probabilistic approximations defined in Sect. 5 are global (they are defined using characteristic sets). In this section we are going to discuss local probabilistic approximations, defined using attribute-value pairs.

Let *X* be any subset of the set *U* of all cases. Let $B \subseteq A$. In general, *X* is not a *B*-definable set, locally or globally.

A *complete B-local probabilistic approximation* of the set *X* with the parameter α, $0 < \alpha \leq 1$, denoted by $appr^{complete}_{\alpha}(X)$, is defined as follows

$$\cup\{[T] | T \text{ is a } B-\text{complex of } X, Pr(X \mid [T]) \geq \alpha\} \tag{20}$$

Complete A-local probabilistic approximations will be called *complete local probabilistic approximations*. For Table 3, the set of all possible blocks of *B*-complexes, where $B \subseteq A$, is the union of the set of all attribute-values blocks and the following set {{1, 4}, {1, 4, 6}, {2, 4, 8}, {4} {4, 5}, {4, 5, 8}, {5}, {5, 7}, {6}}. For Table 3, all distinct complete local probabilistic approximations for the concept [(Trip, yes)] are as follows:

$appr^{complete}_{0.75}(\{1, 2, 3, 4, 5, 6\}) = U$

$appr^{complete}_{0.8}(\{1, 2, 3, 4, 5, 6\}) = \{1, 2, 3, 4, 5, 6, 8\}$

$appr^{complete}_{1}(\{1, 2, 3, 4, 5, 6\}) = \{1, 3, 4, 5, 6\}$

For Table 3, all distinct complete local probabilistic approximations for the concept [(Trip, no)] are as follows:

$appr_{0.2}^{complete}(\{7,8\}) = U$ $\qquad appr_{0.5}^{complete}(\{7,8\}) = \{4,5,7,8\}$

$appr_{0.25}^{complete}(\{7,8\}) = appr_{0.333}^{complete}(\{7,8\})$ $\qquad appr_{0.667}^{complete}(\{7,8\}) = \emptyset$

$= \{2,3,4,5,6,7,8\}$

Note that the set {4, 5, 7, 8}, equal to $appr_{0.5}^{complete}(\{7,8\})$, is not listed in the set of all subset or concept (global) probabilistic approximations.

For a concept X and α, computing $appr_{\alpha}^{complete}(X)$, is a problem of exponential complexity since B-complexes T may contain attribute-value pairs using all possible subsets B of the set A of all attributes. Due to this computational complexity, yet another local probabilistic approximations, called *MLEM2 local probabilistic approximations*, were discussed in Clark et al. (2012). These approximations are computed by using a similar approach as used in the modified MLEM2 algorithm (Grzymala-Busse and Rzasa 2010) for rule induction. The MLEM2 algorithm is of polynomial time complexity.

7 Conclusions

Recently, many experiments on global and local probabilistic approximations were conducted with the same objective: an experimental comparison of usefulness of probabilistic approximations for data mining (Clark and Grzymala-Busse 2011, 2012). In many cases proper probabilistic approximations, i.e., probabilistic approximations different from lower and upper approximations of the same type, are not better or worse than corresponding lower or upper approximations. On the other hand, for some data sets, proper probabilistic approximations are significantly better, though for other data sets, such approximations are significantly worse. Thus, the general conclusion is that for any specific data set, with given type of missing attribute values, probabilistic approximations are worth trying.

References

Clark PG, Grzymala-Busse JW (2011) Experiments on probabilistic approximations. In: Proceedings of IEEE international conference on granular computing, pp 144–149

Clark PG, Grzymala-Busse JW (2012) Experiments using three probabilistic approximations for rule induction from incomplete data sets. In: Proceedings of European conference on data mining, pp 72–78

Clark PG, Grzymala-Busse JW, Kuehnhausen M (2012) Local probabilistic approximations for incomplete data. In: Proceedings of 20th international symposium on methodologies for intelligent systems, pp 93–98

Grzymala-Busse JW (2003) Rough set strategies to data with missing attribute values. In: Proceedings of 3rd international conference on data mining, pp 56–63

Grzymala-Busse JW (2004) Data with missing attribute values: Generalization of indiscernibility relation and rule induction. Trans Rough Sets 1:78–95

Grzymala-Busse JW (2011) Generalized parameterized approximations. In: Proceedings of 6th international conference on rough sets and knowledge technology, pp 136–145
Grzymala-Busse JW (2013) Generalized probabilistic approximations. In: Proceedings of 6th international conference on human system interaction, pp 13–17
Grzymala-Busse JW, Rzasa W (2010) A local version of the MLEM2 algorithm for rule induction. Fundamenta Informaticae 100:99–116
Kryszkiewicz M (1995) Rough set approach to incomplete information systems. In: Proceedings of 2nd annual joint conference on information sciences, pp 194–197
Slowinski R, Vanderpooten D (2000) A generalized definition of rough approximations based on similarity. IEEE Trans Knowl Data Eng 12:331–336
Stefanowski J, Tsoukias A (1999) On the extension of rough sets under incomplete information. In: Proceedings of 7th international workshop on new directions in rough sets, data mining, and granular-soft computing, pp 73–81
Yao YY (1998) Relational interpretations of neighborhood operators and rough set approximation operators. Inf Sci 111:239–259
Yao YY (2007) Decision-theoretic rough set models. In: Proceedings of 2nd international conference on rough sets and knowledge technology, pp 1–12
Yao YY, Wong SKM (1992) A decision theoretic framework for approximate concepts. J Man–Mach Stud 37:793–809

Computer-Aided Relative Assessment of Complex Effects of Decisions

J. L. Kulikowski

Abstract The chapter presents a definition of composite decision making problems as problems consisting in choosing desired actions in the case when they are connected with additional undesired effects. It is assumed that the undesired effects cannot be evaluated but by their pair-wise relative comparison. There are defined algebraic tools which can be effectively used to solution of composite decision making problem: the operations of balancing and extension of preferences and of direct balancing, direct extension and structural product of matrices describing the corresponding semi-ordering relations. Application of the algebraic tools to solution of several types of composite decision making problems is illustrated by numerical examples. Conclusions are formulated in the last section of the chapter.

1 Introduction

A typical and the simplest decision making problem arising in various application areas can be formulated as follows:

> It is known an initial state $\boldsymbol{x}$, $\boldsymbol{x} \in X$, of a social, physical, biological, etc. object and a set U of admissible actions u, transforming $\boldsymbol{x}$ into a new (more desired) object's state, u: $X \to X$; it is also given a function r: $X \times X \to R^+$, R^+ denoting a real non-negative half-axis, assigning a numerical cost $r[\boldsymbol{x}, u(\boldsymbol{x})]$ to the action u applied to $\boldsymbol{x}$; find a decision $u^*, u^* \in U$, minimizing the cost.

A solution of this problem can be provided by any suitably chosen mathematical optimization technique (Michalewicz and Fogel 2004; Cormen et al. 1994). However, the above-formulated decision making problem is not quite adequate to real situations. We are faced sometimes with problems like this: a certain patient suffers from a disease x; it is possible to prescribe him some of drugs: $u_1, u_2, \ldots, u_k$. However, the

J. L. Kulikowski (✉)
Nalecz Institute of Biocybernetics and Biomedical Engineering PAS, Warszawa, Poland
e-mail: jkulikowski@ibib.waw.pl

Z. S. Hippe et al. (eds.), *Issues and Challenges in Artificial Intelligence*,
Studies in Computational Intelligence 559, DOI: 10.1007/978-3-319-06883-1_9,

prices of drugs are different, their effectiveness can only relatively be assessed, some of them cause side effects. The problem consists in choosing a proper drug for the given patient. More generally, in a situation x it is desired to undertake an action u taking into account that: (a) the action may result also in several inevitable side effects $z^{(i)}(x, u), i = 1, 2, 3, \ldots, I$; (b) some resulting effects are desirable, some other are not, (c) the certainty levels of the side-effects are different and can only relatively be assessed, (d) the profits of the desirable effects and the costs of the undesired ones can be comparatively assessed rather than exactly calculated. This type of problems will be called below the *composite decision making* (*CDM*) problems. The *CDM* problems are related to those of multi-aspect optimization of decisions (Peschel and Riedel 1976); however, in the below presented case no numerical evaluation of decisions by vector cost function is assumed. In this sense, the *CDM* is a sort of decision making under uncertainty (Russel and Norvig 2003) close to a natural human decision making. Attempts to its solution lead to the models based on qualitative deliberations rather than on exact optimization calculus; as such, they seem to be close to a sort of intuitive thinking. On the other hand, our level of understanding the natural thinking mechanisms is still very low, despite the fact that various approaches to their description were undertaken (Edward de Bono 1969; Penrose 1994; Dennet 1996; Myers 2004). As the computer-based human decisions supporting systems are introduced to numerous application areas, the problem of natural thinking imitation by computers becomes more important. In this chapter an idea is presented that its solution can be based on some adequately chosen, easy to be implemented in computer systems, algebraic tools for description of the compositions of semi-ordering relations in the decision space U. The compositions should be defined so as to find a trade-off between the profitable and the unprofitable effects of decisions. A solution of the *CDM* problems proposed in the chapter have been inspired by the concept of relative assessment of statements originally formulated in Hempel (1937), as well as by some concepts of extended algebra of relations given in Kulikowski (1992). In Sect. 2 formal backgrounds of an approach to the solution of the *CDM* problems are presented. There are defined the operations of balancing and extension of the preferences imposed on the pairs of compared elements, as well as based on them algebraic operations on the matrices describing the semi-ordering relations. Application of the introduced theoretical tools to solution of several types of *CDM* problems is illustrated by examples. Concluding remarks are presented in Sect. 3.

2 Formal Backgrounds of the *CDM* Problems Solution

2.1 The CDM Problems Characteristics

A formal *CDM* problem is usually connected with a single or a collection of objects of a given universe. However, any collective *CDM* problem, at least formally, can be reduced to a single-object problem by a suitable extension of the set

X of the objects' states so as to represent the states of the considered collections of objects. Hence, below the single-object *CDM* problems only will be considered. For any such problem the following elements should be specified:

1. a set X of the objects' states;
2. a set U of admissible objects' states transformations;
3. a set R of the costs of the objects' states transformations;
4. a set Z of possible additional effects of the objects' states transformations;
5. a relation W weighting the effects of the elements of Z;
6. a relation P describing the relative possibility that the elements of Z will occur when the elements of X and U are given.

The *CDM* problems can thus be divided into several groups, according the assumptions concerning the set Z and the relations W and P:

1. The set Z consists of a single or of several elements;
2. The relation W is described by a numerical weight function or defined as a more general ordering relation;
3. The relation P is described by a (in particular—uniform) probability distribution or defined as a more general ordering relation.

This leads to at least twelve categories of *CDM* problems. The differences between them consist in various types of uncertainty of the effects caused by the undertaken actions. However, in all cases problem solution consists in looking for maximal elements induced by specific types of their ordering. This leads to the algebra of relations as a suitable formal tool for their description. For this purpose, below the compositions of the equivalence and partial ordering relations [see e.g. Kulikowski (1992), Rudeanu (2012)] will be considered.

2.2 Formal Representation of Ordering Relations

It will be taken into consideration a non-empty set Ξ. Its Cartesian square

$$\Xi^2 = \Xi \times \Xi \tag{1}$$

is defined as a set of all possible pairs $[\xi_i, \xi_j]$ of the elements of Ξ. Any its subset $\rho \subseteq \Xi^2$ is called a binary relation defined on Ξ. The pairs $[\xi_i, \xi_j]$ satisfying the relation ρ are called syndromes of the relation. The following properties are usually taken into account in description of binary relations.

Definition 1 A binary relation ρ is called:

(a) *reflexive* if $[\xi_i, \xi_i] \in \rho$ holds for all $\xi_i \in \Xi$;
(b) *irreflexive* if $[\xi_i, \xi_i] \in \rho$ does not hold for any $\xi_i \in \Xi$;
(c) *symmetrical* if for any $\xi_i, \xi_j \in \Xi$ from $[\xi_i, \xi_j] \in \rho$ it follows $[\xi_j, \xi_i] \in \rho$;
(d) *anti-symmetrical* if for any $\xi_i, \xi_j \in \Xi$ from $[\xi_i, \xi_j] \in \rho$ it follows that $[\xi_j, \xi_i] \in \rho$ does not hold;

(e) transitive, if for any $\xi_i, \xi_j, \xi_k \in \Xi$ from $[\xi_i, \xi_j] \in \rho$ and $[\xi_j, \xi_i] \in \rho$ it follows that $[\xi_i, \xi_k] \in \rho$;
(f) anti-transitive if for any $\xi_i, \xi_j, \xi_k \in \Xi$ from $[\xi_i, \xi_j] \in \rho$ and $[\xi_j, \xi_k] \in \rho$ it follows that $[\xi_i, \xi_k] \in \rho$ does not hold. •

Symbol • is used for ending the definitions.

Definition 2 A binary reflexive, symmetrical and transitive relation is called *equivalence*; the notation $\xi_i \approx \xi_j$ (read: ξ_i and ξ_j are equivalent) will be used for the pairs of equivalent elements of Ξ. •

Definition 3 A binary reflexive and transitive relation ρ described in Ξ is called a *quasi-ordering*. The fact that a pair $[\xi_i, \xi_j]$ satisfies the quasi-ordering will be denoted by $\xi_i \preccurlyeq \xi_j$ (read: ξ_j is weakly preceded by ξ_i) or, equivalently, by $\xi_j \succcurlyeq \xi_i$. •

In a given set Ξ many different relations can be defined. If several of them are equivalences or orderings the denoting them symbols will be marked by additional subscripts; for example, $\approx_a$ and $\approx_b$ denote two different relations of equivalence defined on the same set Ξ. Moreover, due to the fact that binary relations are defined as subsets of the Cartesian product Ξ^2, the concepts of set algebra (Rudeanu 2012) can be applied to the family F of all binary relations (including an empty relation $\emptyset$ and a trivial relation Ξ^2) that can be described on Ξ^2. Therefore, for any two binary relations ρ', ρ'' defined on the same set Ξ the following operations can be defined: (a) a negation $\neg\rho'$ of the relation ρ' (as well as a negation $\neg\rho''$ of ρ''), (b) a sum $\rho' \cup \rho''$ of the relations, (c) an intersection $\rho' \cap \rho''$, (d) a difference $\rho' \backslash \rho''$ of the relations (as well as a difference $\rho'' \backslash \rho'$), whose properties are similar to those of the corresponding operations performed on the sets.

In similar way, the notions of set inclusion ($\subseteq$) and proper set inclusion ($\subset$) can be used to define a sub-relation ρ', $(\rho' \subseteq \rho)$ and a proper sub-relation ρ', $(\rho' \subset \rho)$ of the relation ρ.

Definition 3 admits existence of some pairs $[\xi_i, \xi_j]$ of (not only identical) elements satisfying both $\xi_i \preccurlyeq \xi_j$ and $\xi_i \succcurlyeq \xi_j$. All such pairs of elements satisfy the conditions of Definition 2 and, thus, they constitute relations of equivalence.

Definition 4 A binary reflexive, anti-symmetrical and transitive relation ρ described in Ξ is called a *semi-ordering*. The fact that a pair $[\xi_i, \xi_j]$ satisfies the semi-ordering will be denoted by $\xi_i \prec \xi_j$ (read: ξ_j is strongly preceded by ξ_i) or, equivalently, by $\xi_j \succ \xi_i$. •

Definition 5 A quasi-ordering ρ is called a *partial quasi-ordering* if its negation $\neg\rho$ is a non-empty relation. •

Definition 5 can be extended on partial semi-ordering, as well. In both cases partial ordering admits existence of some pairs of elements among which no order has been established. However, $\neg\rho$ constitutes an irreflexive and symmetrical relation of incomparability of some pairs of elements.

Example 1 Various types of ordering relations are illustrated in Fig. 1 by directed graphs. The nodes are assigned there to the elements while the arcs indicate the order assigned to the pairs of elements.

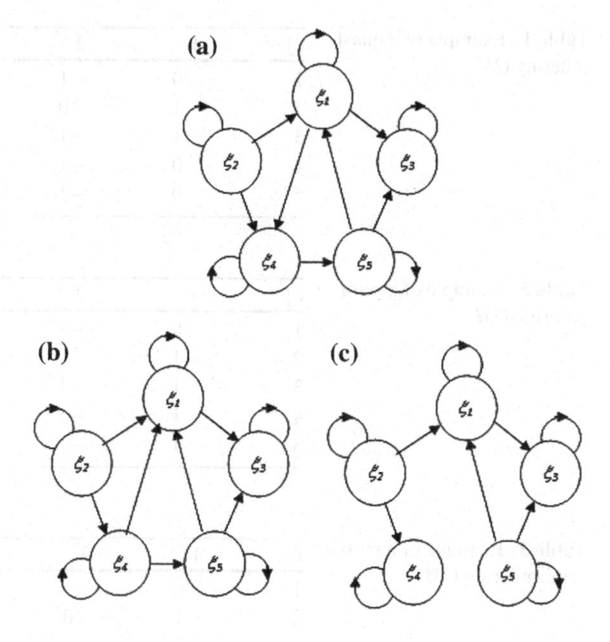

Fig. 1 Directed graphs representing different types of ordering: **a** quasi-ordering, **b** semi-ordering, **c** partial semi-ordering

Figure 1a represents a quasi-ordering (Definition 3): any pair of nodes is connected by one or more directed paths (sequences of uniformly oriented arcs). A directed path [ξ_1, ξ_4, ξ_5] constitutes a closed cycle; its elements satisfy the condition of equivalence (Definition 2). Figure 1b corresponds to a semi-ordering (Definition 4): any pair of nodes is connected by directed path or paths. Figure 1c illustrates a partial semi-ordering (a particular case of Definition 5): some pairs of nodes can be connected by directed paths; however, some other ones (e.g. ξ_2 and ξ_5) by no path can be connected. Graphical representation of ordering relations are useful in human reasoning; however, for computer analysis an algebraic representation seems to be more convenient. For this purpose, a symbolic matrix representation of orderings is proposed.

Definition 6 For a given ordering relation ρ described on a finite set Ξ of N elements a $N \times N$ square matrix $\boldsymbol{W}_\rho = [w_{i,j}]$ such that: $w_{i,j} \in \{1, -1, 0, -0\}$ and (a) $w_{i,j} = 1$ if $\xi_i \prec \xi_j$ and not $\xi_i \succ \xi_j$, (b) $w_{i,j} = -1$ if $\xi_i \succ \xi_j$ and not $\xi_i \prec \xi_j$, (c) $w_{i,j} = 0$ when both $\xi_i \prec \xi_j$ and $\xi_i \succ \xi_j$ hold, (d) $w_{i,j} = -0$ if neither $\xi_i \prec \xi$, nor $\xi_i \succ \xi_j$ holds, will be called an *ordering matrix (OM)* induced by ρ. The pairs [ξ_i, ξ_j] with the corresponding, assigned to them values $w_{i,j}$, will be called *preferences*. •

The elements $w_{i,j}$ of *OM*s should not be interpreted as numerical values, but as symbolic denotations of some statements. However, using numerical denotations is convenient from the computer implementation of *OM*s point of view. For the orderings presented in Fig. 1 the corresponding *OM*s have the following form (Tables 1, 2, 3):

Remark 1 The following structural properties of *OMs* can be remarked.

Table 1 Example of a quasi-ordering *OM*

i\j	1	2	3	4	5
1	0	−1	1	0	0
2	1	0	1	1	1
3	−1	−1	0	−1	−1
4	0	−1	1	0	0
5	0	−1	1	0	0

Table 2 Example of a semi-ordering *OM*

i\j	1	2	3	4	5
1	0	−1	1	−1	−1
2	1	0	1	1	1
3	−1	−1	0	−1	−1
4	1	−1	1	0	1
5	1	−1	1	−1	0

Table 3 Example of a partial semi-ordering *OM*

i\j	1	2	3	4	5
1	0	1	1	−0	−1
2	1	0	1	1	1
3	−1	−1	0	−0	−1
4	−0	−1	−0	0	−0
5	1	−1	1	−0	0

Table 4 Operation of order balancing

&	1	−1	0	−0
1	1	0	0	1
−1	0	−1	0	−1
0	0	0	0	0
−0	1	−1	0	−0

1. The *OM*s are *quasi-symmetrical*, i.e. anti-symmetrical with respect to their 1 and −1 elements and symmetrical with respect to the 0 and −0 elements.
2. Main diagonals of the *OM*s are filled with null-elements 0. However, in semi-ordering and partial semi-ordering *OM*s no other 0 elements occur.
3. In the partial semi-ordering *OM* some −0 elements occur.

For the preferences described by *OM*s two algebraic operations will be introduced.

Definition 7 For the elements {1, −1, 0, −0} the operations of *balancing* & and *extension* ↑ of preferences are defined by Tables 4 and 5:

The following properties of the above-defined operations can be proven by a direct substitution of any admissible values of the variables p, q, r. •

Table 5 Operation of order extension

$\uparrow$	1	−1	0	−0
1	1	−0	1	−0
−1	−0	−1	−1	−0
0	1	−1	0	−0
−0	−0	−0	−0	−0

Corollary *If* $p, q, r \in \{1, -1, 0, -0\}$ *then:*

$$p\&q \equiv q\&p;\ p \uparrow q \equiv q \uparrow p;$$
$$p\&q\&r \equiv p\&(q\&r) \equiv (p\&q)\&r;$$
$$p \uparrow q \uparrow r \equiv p \uparrow (q \uparrow r) \equiv (p \uparrow q) \uparrow r;$$
$$(p\&q) \uparrow r \equiv (p \uparrow r)\&(q \uparrow r)$$

•

The operations of balancing and extension of preferences play a substantial role in *OM* construction. They can be extended on the *OM*s in the following way.

Definition 8 Let $\boldsymbol{W}' = \left[w'_{i,j}\right], \boldsymbol{W}'' = \left[w''_{i,j}\right]$ be two *OM*s of the same size $N \times N$. Then:

(a) a matrix $\boldsymbol{W} = \boldsymbol{W}'\&\boldsymbol{W}''$ such that

$$w_{i,j} = w'_{i,j}\& w''_{i,j} \tag{2}$$

will be called a *direct balance* of matrices $\boldsymbol{W}'$ and $\boldsymbol{W}''$;

(b) a matrix $W = W' \uparrow W''$ such that

$$w_{i,j} = w'_{i,j} \uparrow w''_{i,j} \tag{3}$$

will be called a *direct extension* of matrices $\boldsymbol{W}'$ and $\boldsymbol{W}''$. •

Remark 2 The properties of the operations & and ↑ described in the Corollary hold also in the case of the operations of direct balance and extension of *OM*s.

An order is usually established on the basis of experts' proposals of preferences between selected pairs of the elements of a considered set. Primarily, such indications presented in the form of a matrix $\boldsymbol{m}^{(0)}$ may not satisfy the conditions of the Definitions 4 and 5. This is why at the first step of order establishing the structural requirements of semi-ordering or partial semi-ordering should be imposed upon $\boldsymbol{m}^{(0)}$ and the proposed initial version of *OM* takes the form of a matrix $\boldsymbol{m}^{(1)}$. In fact, the elements of $\boldsymbol{m}^{(1)}$ represent the directed paths consisting only of single preferences—edges in the graph, while *OM* should represent the paths of any (up to $N - 1$) preferences. For finding out all directed paths a notion of *structural product* of *OM*s will be used. This concept is an adaptation of the concept of *Cartesian product of matrices* used to finding all paths in directed graphs (Kulikowski 1986).

Definition 9 Let $\boldsymbol{W}' = \left[w'_{i,j}\right]$, $\boldsymbol{W}'' = \left[w''_{p,q}\right]$ be two *OM*s of the same size $N \times N$. Then a matrix

$$\boldsymbol{W} = \left[w_{i,q}\right] = \boldsymbol{W}' \otimes \boldsymbol{W}'' \tag{4}$$

such that

$$w_{i,q} = \&_{\{j\}}\left(w'_{i,j} \uparrow w''_{j,q}\right) \tag{5}$$

and $\&_{\{j\}}$ denotes balancing of the preferences indexed respectively by j, will be called a *structural product* of the *OM*s $\boldsymbol{W}'$ and $\boldsymbol{W}''$. •

The consecutive approximations of the desired *OM* can be calculated according to the formula:

$$\boldsymbol{m}^{(k)} = (\boldsymbol{m}^{(k-1)} \otimes \boldsymbol{m}^{(1)})\,\&\,\boldsymbol{m}^{(k-1)} \tag{6}$$

for $k = 2, 3, \ldots, N-1$, $\boldsymbol{m}^{(N-1)} \equiv OM$. However, at each step of calculating the *OM* its correctness consisting in satisfying the conditions of Definition 6 should be proven. In particular, any pair of preferences 1 or -1 located symmetrically with respect to the main diagonal indicates on existence of a closed cycle of preferences. In such case a correction to the given matrix $\boldsymbol{m}^{(k)}$ should be introduced by one of two alternative ways. First, existence of the closed cycle can be approved; the preferences the given cycle consists of should be replaced by 0s and the process of calculating $\boldsymbol{m}^{(k)}$ can be continued. Otherwise, the cycle should be broken by replacing one of its preferences by –0; the calculations should then start from the beginning with a modified $\boldsymbol{m}^{(0)}$. As a result, the corrected matrix $\boldsymbol{m}^{(k)}$ represents a balance of all directed paths consisting of at most k segments serially connected by the operation of extension and, possibly, of some closed cycles consisting of mutually equivalent elements.
The following notion of *maximal elements* in a semi-ordered set plays a basic role in establishing comparability of actions.

Definition 10 In a finite semi-ordered set Ξ we call *maximal* each its element ξ such that for no other element ξ_i it is $\xi \prec \xi_i$, excepting the case when $\xi_i \approx \xi$. •

Remark 3 The following properties of *OM*s should be remarked:

1. If i-th row of *OM* excepting the $w_{ii} = 0$ consists of -0 preferences only then it cannot be established whether ξ_i is a maximum or it is not.
2. If all preferences of a row of *OM* equal 0 then all other preferences of *OM* equal 0 and all elements of the set Ξ constitute a set of mutually equivalent maximal elements.
3. If i-th row of *OM* contains one or more -1 preferences and no 1 preference then ξ_i is a maximal element.
4. In the case mentioned in (3) if the i-th row besides $w_{i,i} = 0$ contains any other preferences equal 0 then ξ_i and all corresponding elements ξ_j constitute a subset of mutually equivalent maximal elements of Ξ.

5. If i-th row of OM contains both 1 and -1 preferences then ξ_i can not be a maximal element.
6. Definition 10 can be reversed by substituting the symbols $\prec$ by $\succ$ and vice versa; in such case the *minimal elements* of Ξ will be defined.
7. If in the remarks (1)–(5): (a) all notations "*maximal*" are changed by "*minimal*" and vice versa and (b) all notations 1 are changed by -1 and vice versa then the remarks hold for minimal elements of Ξ.

2.3 Comparability of Actions

Formal tools described in the former section can be used to solution of *CDM* problems. Various types of such problems have been specified in Sect. 2.1. Below, our attention will be focused on the problems characterized by non-numerical, relative assessment of actions and caused by them additional effects.

Example 2 Let us assume that it is given a set $U = \{u_1, u_2, \ldots, u_5\}$ of possible actions leading to a certain goal. Each action causes additionally some positive effects (profits). However, they cannot be evaluated numerically. Instead of this, some preferences have been assigned to them by experts. The set of preferences is as follows:

$$\xi_1 \prec \xi_4, \xi_2 \prec \xi_5, \xi_3 \prec \xi_2, \xi_3 \prec \xi_4,$$

The preferences can also be expressed by a matrix:

$$m^{(0)} = \begin{bmatrix} - & - & - & 1 & - \\ - & - & - & - & 1 \\ - & 1 & - & 1 & - \\ - & - & - & - & - \\ - & - & - & - & - \end{bmatrix}$$

most elements of this matrix being undefined. However, they can be replaced by strongly defined elements according to the indications of Definition 6. This leads to the following, first approximation of OM with diagonal elements equal 0, other lacking elements equal -0 and the OM's quasi-symmetry requirements being taken into consideration:

$$m^{(1)} = \begin{bmatrix} 0 & -0 & -0 & 1 & -0 \\ -0 & 0 & -1 & -0 & 1 \\ -0 & 1 & 0 & 1 & 1 \\ -1 & -0 & -1 & 0 & -0 \\ -0 & -1 & -1 & -0 & 0 \end{bmatrix}$$

For description of the corresponding semi-ordering relation, first, according to the Definition 9, the structural product $m^{(1)} \otimes m^{(1)}$ should be calculated. We illustrate the way of calculations by a single element $w_{3,5}$; according to (3) it is:

$$\begin{aligned} w_{3,5} &= (-0 \uparrow -0) \,\&\, (1 \uparrow 1) \,\&\, (0 \uparrow -0) \,\&\, (1 \uparrow -0) \,\&\, (-0 \uparrow 0) \\ &= (-0) \& (1) \,\&\, (-0) \& (-0) \& (-0) = 1 \end{aligned}$$

Table 6 Assumed costs of actions

Action	u_1	u_2	u_3	u_4	u_5
Cost	12	15	10	12	9

This element has thus changed its original value from −0 to 1, while all other elements remain unchanged. Hence, the matrix takes the form:

$$\boldsymbol{m}^{(1)} \otimes \boldsymbol{m}^{(1)} = \begin{bmatrix} 0 & -0 & -0 & 1 & -0 \\ -0 & 0 & -1 & -0 & 1 \\ -0 & 1 & 0 & 1 & 1 \\ -1 & -0 & -1 & 0 & -0 \\ -0 & -1 & -1 & -0 & 0 \end{bmatrix}$$

Balancing this matrix with $m^{(1)}$, according to (4), does not change any of its elements. Hence, in this case is also the final form of *OM*; we denote the result by $(\boldsymbol{m}^{(1)} \otimes \boldsymbol{m}^{(1)})\,\&\,\boldsymbol{m}^{(1)} = \boldsymbol{m}^{(2)} = \boldsymbol{W}$.

Example 3 Let $U = (u_1, u_2, \ldots, u_5)$ denote a set of possible actions leading to a desired final state. Assumed numerical costs r of the actions are given by Table 6.

Of course, the minimal-cost action in this case can be found directly. However, for multi-aspect relative assessment of actions exact numerical costs are not necessary and they can be replaced by the below-given *OM*:

$$\boldsymbol{R} = \begin{bmatrix} 0 & 1 & -1 & 0 & -1 \\ -1 & 0 & -1 & -1 & -1 \\ 1 & 1 & 0 & 1 & -1 \\ 0 & 1 & -1 & 0 & -1 \\ 1 & 1 & 1 & 1 & 0 \end{bmatrix}$$

Let us remark that this, based on numerical costs *OM*, describes a semi-ordering without undefined preferences (contains no −0 elements). Evidently, minimal cost is r_5 because 5th row of $\boldsymbol{R}$ consists of all non-diagonal preferences equal 1 [see the above-formulated remarks (3), (6) and (7)].

Moreover, it is also assumed that with each action additional undesired effect is connected. It is given in the form of the *OM* of preferences suggested by experts, given by the MATRIX $\boldsymbol{W}$ in Example 2. The semi-orderings described by $\boldsymbol{W}$ and $\boldsymbol{R}$ are evidently different. It thus arises the question of their *harmonization* before finding the minimal element (the best action). It is assumed that the preference of any pair of actions should be obtained as a balance of preferences of this pair suggested by two semi-ordering relations. Therefore, it will be done by the operation of balancing $\boldsymbol{W}$ and $\boldsymbol{R}$ which leads to the following *OM*:

$$\boldsymbol{W}' = \boldsymbol{R}\&\boldsymbol{W} = [r_{i,j}\&w_{i,j}] \tag{5}$$

where $r_{i,j}$, $w_{i,j}$ are, respectively, the components of the matrices $\boldsymbol{R}$ and $\boldsymbol{W}$.

The balanced matrix is given below.

$$W' = \begin{bmatrix} 0 & 1 & 0 & 0 & -1 \\ -1 & 0 & -1 & 1 & -0 \\ 1 & 1 & 0 & 1 & -0 \\ 0 & -1 & -1 & 0 & -1 \\ 1 & -0 & -0 & 1 & 0 \end{bmatrix}$$

Its 2nd row containing two (-1)s and no 1 preference corresponds to a single maximal element and the 3rd and 5th rows containing some 1 and no -1 preferences correspond to the minimal elements of W'.
The result indicates that u_3 and u_5 can be recommended as *the best solutions* of the task. However, the solutions are *mutually incomparable* because u_3 is better than u_5 from the cost from of view while u_5 is better than u_3 from the point of view of other undesired effects of the actions.

Example 4 Like in Example 3, it is considered a set $U = (u_1, u_2, \ldots, u_5)$ of admissible actions and assigned to them preferences induced by costs (matrix $\boldsymbol{R}$) and undesired additional effects (matrix $\boldsymbol{W}$). Moreover, it is assumed that the additional effects may occur but with some certainty levels. In a strong formulation of the problem it might be assumed that some probabilities have been to them assigned. In such case, the numerical values of probabilities lead to a matrix of preferences in similar way as it was shown above by construction of $\boldsymbol{W'}$ on the basis of Table 6. However, if the probabilities are unknown, the certainty levels of arising additional effects can be roughly characterized by preferences indicated by experts. For this purpose, it can be established a qualitative scale, like:

$$\textit{never} \prec \textit{rarely} \prec \textit{sometimes} \prec \textit{often} \prec \textit{very often} \prec \textit{usually} \prec \textit{always}$$

However, intuitive sense rather than numerical values (probabilities, membership levels, etc.) are assigned to the terms of this scale. Let it be assumed that the undesired effects of decisions are weighted as follows:

$$\xi_1 - \textit{sometimes}, \xi_2 - \textit{usually}, \xi_3 - \textit{often}, \xi_4 - \textit{rarely}, \xi_5 - \textit{often}$$

This leads to the preferences:

$$\xi_1 \prec \xi_2, \xi_1 \prec \xi_3, \xi_1 \succ \xi_4, \xi_1 \prec \xi_5, \xi_2 \succ \xi_3,$$
$$\xi_2 \succ \xi_4, \xi_2 \succ \xi_5, \xi_3 \succ \xi_4, \xi_3 \succ \xi_5, \xi_4 \prec \xi_5$$

which can be presented by the *OM*:

$$\boldsymbol{P} = \begin{bmatrix} 0 & 1 & 1 & -1 & 1 \\ -1 & 0 & -1 & -1 & -1 \\ -1 & 1 & 0 & -1 & -1 \\ 1 & 1 & 1 & 0 & 1 \\ -1 & 1 & 1 & -1 & 0 \end{bmatrix}$$

The weighted effects will be given by a direct product of the *OM*s $\boldsymbol{W}$ and $\boldsymbol{P}$:

$$\boldsymbol{W'} = \boldsymbol{W} \uparrow \boldsymbol{P} = \begin{bmatrix} 0 & -0 & -0 & -0 & 1 \\ -0 & 0 & -1 & -0 & -0 \\ -0 & 1 & 0 & -0 & -0 \\ -0 & -0 & -0 & 0 & -0 \\ -1 & -0 & -0 & -0 & 0 \end{bmatrix}$$

$\boldsymbol{W'}$ contains two preferences: $w'_1 \prec w'_5$ and $w'_3 \prec w'_2$ indicating the actions u_1 and u_3 charged by minimal undesired effects. However, the actions are mutually incomparable, because the possibility of undesired effects occurrence in connection with u_1 is lower then in connection with u_3 but the costs of the effects are unknown: it may happen that the less frequent effects are charged by relatively much higher cost. Rational choosing between w_1 and w_3 needs thus introducing additional information reducing the number of -0 preferences.

Final solution of the *CDM* problem needs calculation of the balance $\boldsymbol{W'}$ **&** $\boldsymbol{R}$ and of its structural square $(\boldsymbol{W'} \,\&\boldsymbol{R}) \otimes (\boldsymbol{W'} \,\&\boldsymbol{R})$ in order to detect all equivalences of the preferences. This leads to the below-given *OM*:

$$\boldsymbol{W'}\&\boldsymbol{R} = \begin{bmatrix} 0 & 1 & 0 & 0 & 0 \\ -1 & 0 & -1 & -1 & -1 \\ 0 & 1 & 0 & 0 & 0 \\ 0 & 1 & 0 & 0 & 0 \\ 0 & 1 & 0 & 0 & 0 \end{bmatrix}$$

It follows from the *OM* that the maximum is given by the element w_2 while the elements w_1, w_3, w_4, and w_5 constitute a class of mutually equivalent minimal elements. The best action should thus chosen from u_1, u_3, u_4, and u_5 taking into account that they are equivalent for various reasons.

3 Discussion and Conclusions

The *CDM* problems can be formulated as semi-ordering of actions indicating the minimal (in the sense of widely defined costs) or maximal (in the sense of widely defined profits) actions. It was shown that finding the trade offs between the costs and profits is possible due to some adequately chosen algebraic tools like, in particular, the operations of balancing and extension of preferences given by experts. Matrix representation of semi-ordering relations by *OM*s makes this approach easy to computer implementation. However, the method has also some shortcomings which should be removed in further works. First, the harmonization of ordering relations based on formula (5) leads sometimes to a too large number of equivalent or mutually incomparable solutions. It seems thus necessary to extend the idea of harmonization and make it more flexible by diversification of the weights of criteria. It seems also necessary to extend the method on choosing more than one action by taking into account their possible synergetic co-existence or, on the other hand, their opposite influence on the final result.

References

Cormen TH, Leiserson CE, Rivest RL (1994) Introduction to algorithms, 13th printing. The MIT, Cambridge

de Bono E (1969) The mechanism of mind. Mica Management Resources Inc., UK

Dennet DC (1996) Kind of minds: towards an understanding of consciousness. http://en.wikipedia.org/wiki/Daniel_Dennett. Accessed 5 Jan 2013

Hempel CG (1937) A purely topological form of non-Aristotelian logic. J Symbolic Logic 2(3)

Kulikowski JL (1986) Outline of the theory of graphs. PWN, Warsaw

Kulikowski JL (1992) Relational approach to structural analysis of images. Mach Graph Vis 1(1/2):299–309

Michalewicz Z, Fogel DB (2004) How to solve it: modern heuristics. Springer, Berlin

Myers DG (2004) Intuition its powers and perils. Yale University Press, New Haven

Penrose R (1994) Shadows of the mind a search for the missing science of consciousness. Oxford University Press, Oxford

Peschel M, Riedel C (1976) Polyoptimierung. VEB Verlag Technik, Berlin

Rudeanu S (2012) Sets and ordered structures. Bentham Science Publishers, Karachi

Russel S, Norvig P (2003) Artificial intelligence: a modern approach. Prentice Hall, Pearson Education Inc, Upper Saddle River

Suitability of BeesyCluster and Mobile Development Platforms in Modern Distributed Workflow Applications for Distributed Data Acquisition and Processing

P. Czarnul

Abstract A concept of a distributed system for acquisition of data by modern mobile devices is proposed in the chapter. The data are preprocessed, subsequent passed and cached to intermediate servers that expose services for fetching the data. Distinct data gathering zones are proposed, either private or public. The services can be combined into a complex workflow processing on top of the BeesyCluster middleware that provides an easy-to-use human-system interface for management of multiple workflow instances. The chapter discusses suitability of various mobile software development approaches and APIs in terms of gathering data from particular sensors. Finally, an exemplary implementation of data acquisition on the modern PhoneGap platform is provided.

1 Introduction

Recent developments and increase in popularity of mobile devices have made it possible to create whole new projects and initiatives with entirely new applications for the society. This includes:

- location-based services in which users gain functions based on their physical location (ME NewsWire 2013),
- personal assistants not only for office tasks but for monitoring health and daily activities such as sports (Bexelius et al. 2010),
- data acquisition with the use of a wide range of sensors installed in such devices (Kang et al. 2007; Han et al. 2007; Maisonneuve et al. 2009),
- remote control of various appliances using mobile devices such as audio equipment or recent DSLR cameras (Ponce 2013).

P. Czarnul (✉)
Department of Computer Architecture, Faculty of Electronics, Telecommunication and Informatics, Gdańsk University of Technology, Gdańsk, Poland
e-mail: pczarnul@eti.pg.gda.pl

Z. S. Hippe et al. (eds.), *Issues and Challenges in Artificial Intelligence*,
Studies in Computational Intelligence 559, DOI: 10.1007/978-3-319-06883-1_10,

While the recent mobile devices feature powerful multi-core processors that make some applications possible to run locally, there is a group of demanding solutions that require computational power of high performance computing (HPC) systems. This, in turn, requires proper infrastructure for efficient and secure coupling of not only HPC and mobile systems but also provision of an easy-to-use interface for the human.

2 Related Work

Distributed systems that gather and process data from multiple sources can be characterized in terms of particular components such as: data acquisition, processing and analysis, integration of data acquisition, HPC processing, visualization and human-system interaction for complex, geographically distributed scenarios.

There are several modern applications that utilize sensors within smartphones. Air pollution can be measured by the Visibility app for Android developed at University of Southern California (Poduri et al. 2009). Mapping of noise in the area can be done using the approach proposed in Maisonneuve et al. (2009). Measurement of physical activity using cell phones is described in Bexelius et al. (2010). Location-based advertising on mobile phones allows to attract nearby customers (ME NewsWire 2013). Analysis of existing APIs and their suitability for data acquisition is shown in Sect. 5.

Processing of data gathered on devices can be performed either locally or sent to a powerful server or an HPC system. The latter might be due to limitations in memory available, CPU performance or large databases against which the data needs to be compared or verified. It should be noted that today's mobile devices, especially tablets and small notebooks, offer impressive performance with multi-core CPUs and GBs of RAM. Nevertheless, offloading computations to a dedicated system gives access to several HPC APIs available on respective systems (Buyya 1999; Kirk and Hwu 2012):

1. NVIDIA CUDA, OpenCL and OpenACC for GPUs, OpenMP used for SMP systems including modern accelerators such as Intel Phi,
2. MPI for traditional multi-process and multithreaded programming on clusters that can include multiple CPUs,
3. hybrid approaches such as MPI + OpenMP, MPI + CUDA, MPI + OpenCL for clusters that feature not only CPUs but also multiple GPUs.

Workflow-based systems for grids (Yu and Buyya 2005) allow integration of data acquisition and computational services into complex scenarios. The author proposed how the features of BeesyCluster's workflow management system (Czarnul 2013a) can be used for integration of mobile devices and HPC systems in short time. It allows flexible system reconfiguration and various applications.

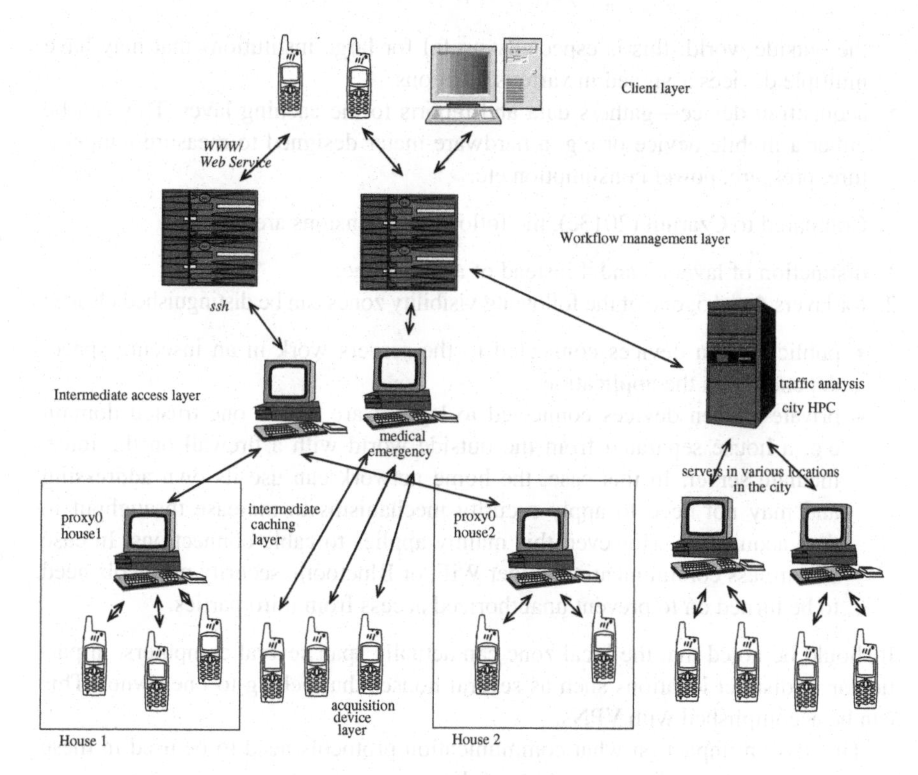

Fig. 1 Proposed multi-tiered architecture for data acquisition and processing

3 Proposed Architecture

Extending the previous work by the author (Czarnul 2013c), Fig. 1 presents a generalized proposed architecture of the system that allows data acquisition from mobile devices and subsequent processing on either mobile devices or HPC systems. Compared to Czarnul (2013c) the context of a whole city, not only control of smart homes is considered. The following layers can be distinguished:

1. client—accessible to the user who needs to define processing flow, start data acquisition and analysis and be able to check status and browse results,
2. workflow management that deals with storage of workflow applications, launching workflow instances based on workflow definitions for various input data, execution, browsing workflow instance statuses and results,
3. intermediate access server—an access server that exposes data gathered by lower level servers and devices that belong to the given institution, organization or home,
4. intermediate caching layer (optional)—a layer that contain servers or computers to which various mobile devices can report data, these are separated from

the outside world; this is especially useful for large institutions that may have multiple devices assigned in various locations,
5. acquisition device—gathers data and reports to the caching layer. This can be either a mobile device or e.g. a hardware meter designed to measure temperature, pressure, power consumption etc.

Compared to Czarnul (2013c), the following extensions are marked:

1. distinction of layers 3 and 4 instead of a single one,
2. for layers 4 and 5, one of the following visibility zones can be distinguished clearly:
    - public—when devices connected to the servers work in an insecure space; depending on the application
    - private—when devices connected to layer 4 are within one trusted domain e.g. a house separated from the outside world with a firewall on the intermediate server. In this case, the home network can use its own addressing and may not need to apply security mechanisms to increase throughput in data acquisition. However, this mainly applies to cable connections. In case of wireless communication, either WiFi or Bluetooth, security protocols need to be turned on to prevent unauthorized access from third parties.

It should be noted that the local zone can actually span several computers, in particular in distinct locations such as several houses that belong to one owner. This can be accomplished with VPNs.

This has an impact on what communication protocols need to be used in these cases especially regarding encryption of data.

Each acquisition device can work in one of two modes:

- client—when it is sending data to one or more of servers,
- server—in which case it can be contacted by one or more of the local servers using e.g. HTTP or Bluetooth and can respond with data sets obtained from the connected sensors.

4 Implementation of the Workflow Management Layer Using BeesyCluster

BeesyCluster (Czarnul et al. 2005) is a middleware that allows to model and deploy the environment shown in Fig. 1. For that purpose, two layers within BeesyCluster are used:

1. core—within this layer users can set up accounts in BeesyCluster and assign system accounts on various clusters, servers and workstations to the BeesyCluster account. Users can publish applications from those accounts as services that are then available on the BeesyCluster platform and can be used within workflow applications.

2. workflow management—includes workflow scheduling, execution using software agents or Java EE servers (Czarnul 2013a), workflow application monitoring and instance management, results' visualization.

Depending on the institution, various access services can be set up at layer 3:

- public: e.g. reporting of locations of inhabitants of the city to detect traffic, offer more frequent city transportation etc.
- private: a server installed within a smart home for reporting of data by devices put in the house in various locations; in some cases, the owner of such a server may give special access passwords to their neighbors and e.g. consider reporting from their devices with a lower priority.

It should be noted that data can be cached and filtered in both intermediate and access servers. The latter may be required to filter out high frequency short-lived observations or compact the size of data.

At the workflow level, each data access service is launched with a set of passwords allowing saving data from the caching server. As outlined in Czarnul (2013c), logging in with a login and a password returns an authenticator that is valid for a particular period of time and allows registration of data. From the point of view of the workflow application, the service works in a streaming mode (Czarnul 2013b) i.e. upon calling always returns data from its local cache.

5 Implementation of Data Acquisition Using Smartphones and Various Platforms

The current Smartphones feature more and more sensors and possibilities to communicate with external devices using protocols such as Bluetooth. This makes them a good choice for data acquisition as their location can be easily changed. Furthermore, the number of Smartphones available is very large and still rapidly growing (Gartner Press Release 2013). This means that the data gathered can also be associated with human population, their location and activities.

5.1 Approaches to Software Development on Mobile Platforms

The current market shares of particular mobile device manufacturers and operating systems are far from being stable. Almost every year, as shown by Gartner reports, important changes are visible. As of the second quarter of 2013, among new devices, the market of new devices is divided among the operating systems as follows: Android with 79 %, iOS 14.2 %, Microsoft Phone 3.3 %, Blackberry 2.7 % and others 0.9 % (Gartner Press Release 2013). This means that programmers wishing to develop for the mobile market face a difficult choice of which programming platform and API to choose:

1. native API—fast and all possible functions included but requires development and maintenance of multiple completely different solutions for 4–5 major operating systems.
2. web based development using the latest HTML5,
3. hybrid approach that couples the web based approach with a possibility to invoke certain native codes through an intermediate layer.

As reported in Hammond et al. (2012), starting with the web-based approach and using the hybrid one when necessary is suggested. Comparison of these approaches in terms of: cost of development, flexibility of modifications, availability of API, performance is shown in Table 1.

5.2 *Support for Data Acquisition on Various Mobile Platforms*

Essentially, modern mobile devices, irrespective of the operating system, offer APIs for various type of sensors from which various types of data can be gathered. Such sensors may include:

1. accelerometer for shake detection,
2. geolocation of the device,
3. camera capture (video and still) for monitoring the environment,
4. audio for detection of breaking glass patterns etc.,
5. light intensity, temperature, humidity, perspiration sensor etc.

Depending on the software development platform presented in Sect. 5.1, ease of access to the data may vary. Generally, the native API offers as good access to the sensor as possible at the cost of development time and cost. Hybrid approaches may be somewhat limited. For instance, the native Android API offers the possibility to gather images from the camera using the default device camera application or accessing the camera directly (PhoneGap Website and Documentation 2013). On the other hand, the default PhoneGap API launches the default camera application only (PhoneGap Website and Documentation 2013). It is also possible to get down to a lower level using plugins (PhoneGap Website and Documentation 2013) but this makes the program more complicated and development time-consuming. Work (Paller 2013) discusses advantages and disadvantages of PhoneGap for data acquisition using sensors. It is concluded that the PhoneGap approach requires more CPU power compared to the native Java approach or an implementation using plugins (9 % compared to 1 %).

Table 1 Comparison of various approaches to mobile application development

Development approach to mobile applications	Cost	Flexibility	Development time	Availability of API	Performance
Native API (Android, iOS, Windows Phone, Blackberry, Symbian)	− High, requires implementations for various operating systems	− Low due to need for modifications in several versions	− Long due to several versions	+ All possible functions available	+ No overhead
Web based	+ Very portable as browsers are available for major platforms	± (almost) Same code for all platforms	+ (very fast prototyping)	− API limited, certain functions (e.g. of HTML5) not available on all platforms, browsers (e.g. WebWorkers)	±
Hybrid (web based + native calls)	± Portable, may require various configurations, various small code parts for different platforms	± (almost) Same code for all platforms, may require writing plugins for some functions	± Very fast, some configuration specific to various platforms	± Some functions limited compared to native APIs	±

Table 2 Easy support to full data access related to the given sensor: *F* (full), *L* (limited)

Sensor	Native API	Hybrid approach (based on PhoneGap)
Accelerometer	F	F
Video/Camera	F (can use default camera application or camera API)	L (uses default camera application)
Audio	F	L (uses default application)
Geolocation	F	F
Compass	F	F
Others	F	L (requires plugins)

Additionally, it should be noted that for the given device, one or the other API might be preferred for data acquisition. Taking the aforementioned example, using the standard camera application might not be suitable for periodic and remote launching as it would require actual picture taking by the user. Consequently, this might require another camera application or accessing lower level APIs using plugins. For instance, a plugin for the temperature sensor is available (TempListener 2010). From this point of view, Table 2 presents preferred APIs for particular sensors based on exemplary APIs for native and hybrid approaches such as Android native API and PhoneGap.

5.3 Processing Locally or on HPC System?

In case of certain sensors and devices, processing can be executed on the device itself while for more complex processing can be easily sent to the HPC system. The latter is easily possible using the well established HTTP which can be handled as shown in Fig. 2 or by using the FileTransfer object available in the PhoneGap environment. It should be noted that communication using XMLHttpRequest can be performed either synchronously or asynchronously.

Since monitoring and reporting can use many sensors and devices at the same time and must not interfere with the main view of the application, multiple threads should be engaged. This is possible through a new construct in HTML5 i.e. WebWorkers that execute given code in the background. However, this requires proper support from the browsers and this can vary. Alternatively, the Javascript setInterval method can be used to launch periodic calls separated by the given delay.

A question arises where analysis of particular types of data should be performed. The author proposes preferences as outlined in Table 3 with notes on particular applications.

```
<html>
  <head> (...)
    <script type="text/javascript" charset="utf-8"
src="cordova.js"></script>
    <script type="text/javascript" charset="utf-8">
    ...
    function observeAccel() { // launched when application starts
        var o = { frequency: 10000 };
        navigator.accelerometer.watchAcceleration(onSuccess, onError,
o);
    }
    function load (notification,auth) {
          var xmlhttp;
          if (window.XMLHttpRequest)
                   {  xmlhttp=new XMLHttpRequest(); } else
                   {// version for Internet Explorer 6
                   xmlhttp=new ActiveXObject("Microsoft.XMLHTTP");
                   }
          xmlhttp.onreadystatechange=function() {
                   if (xmlhttp.readyState==4 && xmlhttp.status==200){
                   document.getElementById("myResp").
                   innerHTML=xmlhttp.responseText;
                   }
          }
          xmlhttp.open("POST","serverURL",true);
          xmlhttp.setRequestHeader("Content-type","application/x-www-
          form-urlencoded");
          xmlhttp.send("auth="+auth+"</br>notif="+notification);
    }
    function onSuccess(acceleration) {
          var element = document.getElementById('accelerometer');
          message='Acceleration in X: '
          +acceleration.x+'<br/>'+'Acceleration in Y: ' +
          acceleration.y +'<br />'+'Acceleration in Z: ' +
          acceleration.z + '<br />' +'When: '       +
          acceleration.timestamp + '<br />';
          element.innerHTML =message;
          load (message,<authenticator>);
    }
    </script>
  </head>
  <body>
    <div id="accelerometer">Wait for accelerometer data...</div>
    Confirmation from server
    <div id="myResp"></div>
  </body>
</html>
```

Fig. 2 Sample PhoneGap application for acceleration watch

5.4 An Example Using the PhoneGap Platform

Figure 2 presents code that uses accelerometer to detect shake on the server side. Compared to the basic template shown in PhoneGap Website and Documentation (2013), the code reports acceleration (passes a previously obtained authenticator) and the server makes a decision whether sufficiently large acceleration has been detected.

Table 3 Recommended location for processing selected sensor data with comments

Sensor	Mobile device	HPC (external) system
Accelerometer	In case of determination of plain changes or gestures (Wang et al. 2012)	If comparison against many patterns or historical data needed
Video/Camera	Selected determination applications such as: Palmprint recognition (Han et al. 2007) License plate recognition (Kang et al. 2007) Shape retrieval and recognition (Kovacs 2012) Logo recognition (Nguyen et al. 2013)	For comparison against large databases of images, finding objects within image etc. e.g. a Google Glass application (Simonite 2013)
Audio	Local voice recognition	Music recognition, speech translation e.g. SpeechTrans (Venables 2013)
Geolocation	Location-based services with database on the device e.g. POI databases in satellite navigation systems etc.	Location-based services with large databases available on the server side

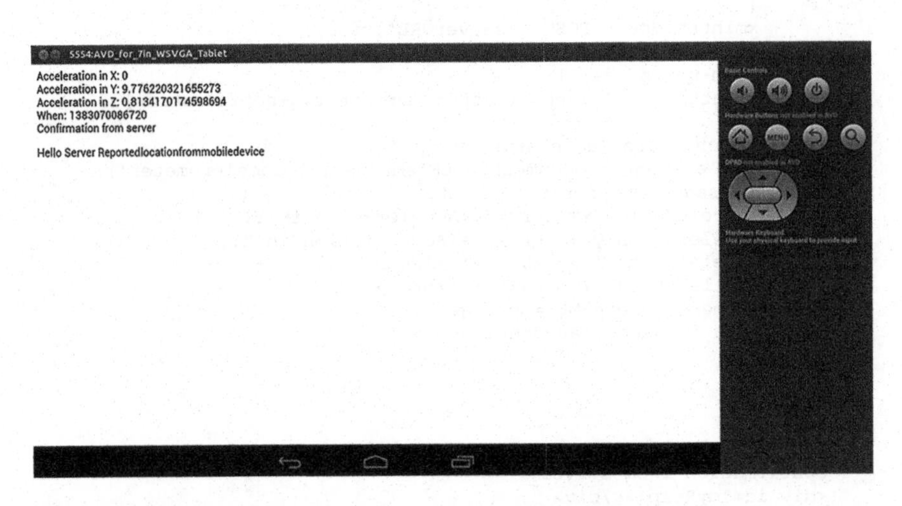

Fig. 3 Screenshot of the PhoneGap application on the Android platform

Furthermore, Fig. 3 shows a simulator of an 7″ tablet running Android with the code shown in Fig. 2 running on it.

6 Conclusions

The chapter presented a proposal of integration of BeesyCluster and mobile applications for distributed, potentially large-scale data acquisition and processing. What is important, software development platforms and APIs for mobile devices were analyzed in terms of suitability for this purpose considering particular sensors available on the mobile device. Furthermore, analysis was performed regarding suitability of data processing on either the mobile device or an HPC system for particular sensors. On the other hand, BeesyCluster was proposed as a middleware that can make use of the data supplied by the mobile devices, route it to services and arrange complex and repeatable processing tasks. An example of handling the acceleration sensor was provided on the modern PhoneGap platform with communication to a server accessible from BeesyCluster. Several practical applications of this setup were described.

Acknowledgments The work was partially performed within grant "Modeling efficiency, reliability and power consumption of multilevel parallel HPC systems using CPUs and GPUs" sponsored by and covered by funds from the National Science Center in Poland based on decision no DEC-2012/07/B/ST6/01516.

References

Bexelius C, Sandin S, Lagerros YT, Forsum E, Litton J (2010) Measures of physical activity using cell phones: validation using criterion methods. J Med Internet Res 12(1):e2

Buyya R (1999) High performance cluster computing, programming and applications. Prentice Hall, Englewood Cliffs

Czarnul P (2013a) Modeling, run-time optimization and execution of distributed workflow applications in the jee-based beesycluster environment. J Supercomput 63:46–71

Czarnul P (2013b) A model, design, and implementation of an efficient multithreaded workflow execution engine with data streaming, caching, and storage constraints. J Supercomput 63(3):919–945

Czarnul P (2013) Design of a distributed system using mobile devices and workflow management for measurement and control of a smart home and health. In: Proceedings of the 6th international conference on human system interaction, pp 184–192

Czarnul P, Bajor M, Fraczak M, Banaszczyk A, Fiszer M, Ramczykowska K (2005) Remote task submission and publishing in beesycluster: security and efficiency of web service interface. In: Lecture notes in computer science, vol 3911, pp 220–227

Gartner Press Release (2013). http://www.gartner.com/newsroom/id/2573415. Accessed 5 Jan 2014

Hammond JS, McNabb K, Coyne S (2012) Building mobile apps? Start with web; move to hybrid—a social computing report. Forrester

Han Y, Tan T, Sun Z, Hao Y (2007) Embedded palmprint recognition system on mobile devices. In: Advances in biometrics. Lecture notes in computer science, vol 4642, pp 1184–1193

Kang JS, Jeong TT, Oh SH, Sung MY (2007) Image streaming and recognition for vehicle location tracking using mobile devices. In: Advances in grid and pervasive computing. Lecture notes in computer science, vol 4459, pp 730–737

Kirk DB, Hwu W (2012) Programming massively parallel processors, second edition: a hands-on approach. Morgan Kaufmann, Los Altos

Kovacs L (2012) Shape retrieval and recognition on mobile devices. In: Computational intelligence for multimedia understanding. Lecture notes in computer science, vol 7252, pp 90–101

Maisonneuve N, Stevens M, Steels L (2009) Measure and map noise pollution with your mobile phone. In: Proceedings of the 27th annual CHI conference on human factors in computing systems, pp 78–82

ME NewsWire (2013) NEARBUY launches its unique shopping location based mobile app in Dubai. http://me-newswire.net/news/7838/en. Accessed 5 Jan 2014

Nguyen PH, Dinh TB, Dinh TB (2013) Local logo recognition system for mobile devices. In: Computational science and its applications. Lecture notes in computer science, vol 7975, pp 558–573

Paller G (2013) Advantages and limitations of PhoneGap for sensor processing. Sfonge Ltd., Droidcon Tunis

PhoneGap Website and Documentation (2013). http://www.phonegap.com. Accessed 5 Jan 2014

Poduri S, Nimkar A, Sukhatme G (2009) Visibility monitoring using mobile phones. http://robotics.usc.edu/mobilesensing/visibility/MobileAirQualitySensing.pdf. Accessed 5 Jan 2014

Ponce D (2013) Control your dSLR remotely with your Smartphone. http://www.ohgizmo.com/2013/07/29/control-your-dslr-remotely-with-your-smartphone/. Accessed 5 Jan 2014

Simonite T (2013) A Google glass app knows what you're looking at. MIT Technol Rev. http://www.technologyreview.com/view/519726/a-google-glass-app-knows-what-youre-looking-at/. Accessed 5 Jan 2014

TempListener (2010). Nitobi Software Inc. IBM Corporation. http://svn.apache.org/repos/asf/incubator/callback/phonegap-android/trunk/framework/src/com/phonegap/TempListener.java. Accessed 5 Jan 2014

Venables M (2013) Review: speechtrans and Google translate at home and on the road. http://www.wired.com/geekdad/2011/03/review-speechtrans-and-google-translate-at-home-and-on-the-road/. Accessed 5 Jan 2014

Wang X, Tarrío P, Metola E, Bernardos AM, Casar JR (2012) Gesture recognition using mobile phone's inertial sensors. In: Distributed computing and artificial intelligence. Advances in intelligent and soft computing, vol 151, pp 173–184

Yu J, Buyya R (2005) A taxonomy of workflow management systems for grid computing. J Grid Comput 3:171–200

Part III
Optimization

Genetic Programming for Interaction Efficient Supporting in Volunteer Computing Systems

J. Balicki, W. Korłub, H. Krawczyk and J. Paluszak

Abstract Volunteer computing systems provide a middleware for interaction between project owners and great number volunteers. In this chapter, a genetic programming paradigm has been proposed to a multi-objective scheduler design for efficient using some resources of volunteer computers via the web. In a studied problem, genetic scheduler can optimize both a workload of a bottleneck computer and cost of system. Genetic programming has been applied for finding the Pareto solutions by applying an immunological procedure. Finally, some numerical experiment outcomes have been discussed.

1 Introduction

In the grid and volunteer computing systems like BOINC or Comcute, some scientific projects are transformed to a set of the calculation tasks that are executed concurrently by volunteer computers with a support of some levels of the middleware modules. A society of scientists can use these systems for extensive distributed calculations in some research projects. The 24-h average performance of the most popular volunteer system BOINC is 8.186 TeraFLOPS.

J. Balicki (✉) · W. Korłub · H. Krawczyk · J. Paluszak
Faculty of Telecommunications, Electronics and Informatics,
Gdansk University of Technology, Gdańsk, Poland
e-mail: balicki@eti.pg.gda.pl

W. Korłub
e-mail: waldemar.korlub@pg.gda.pl

H. Krawczyk
e-mail: henryk.krawczyk@eti.pg.gda.pl

J. Paluszak
e-mail: jpaluszak@gmail.com

Z. S. Hippe et al. (eds.), *Issues and Challenges in Artificial Intelligence*,
Studies in Computational Intelligence 559, DOI: 10.1007/978-3-319-06883-1_11,

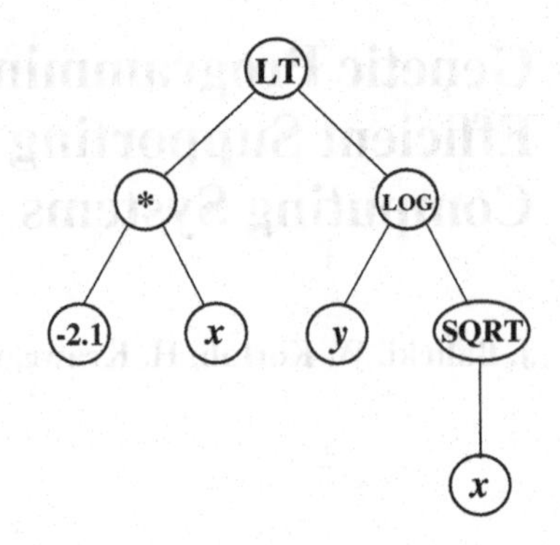

Fig. 1 An example of a parse tree as a chromosome of genetic programming

Moreover, the number of active volunteers can be estimated as 238,412, and also 388,929 computers process data (BOINC 2013).

In the Comcute system, that has been developed at the Gdansk University of Technology, an application for the Collatz hypothesis verification and another one for finding the next Mersenne prime number are applied to prove the intense human interactions, scalability and high performance. The Comcute allows some connecting volunteers to fetch codes and subsequently input data packets. Then some outcomes are returned to a server side. This system permits definition of several tasks and computations for several task instances at the same time.

Genetic programming starts from a goal to be achieved and then it creates an application autonomously without explicitly programming (Koza 1992). To some extent, it replies to the question that has been formulated by Arthur Samuel—a founder of machine learning—"How can computers be made to do what needs to be done, without being told exactly how to do it?" (Samuel 1960). This paradigm uses the principle of selection, crossover and mutation to obtain a population of programs. Genetic programming has been successfully applied to some problems from different fields (Koza et al. 2003). Multi-criterion genetic programming (MGP) can determine the Pareto-optimal solutions (Balicki 2006).

In this chapter, MGP has been proposed to a multi-objective scheduler design for efficient using some resources of volunteer computers. The scheduler optimizes both a workload of a bottleneck computer and the cost of the system. Moreover, genetic programming has been applied for finding the Pareto solutions by development an immunological system based procedure. Finally, some outcomes for numerical experiments have been presented.

2 Genetic Programming Rules

Solutions to several problems have been found by genetic programming (GP) for instances from different areas, e.g. optimal control, planning and sequence induction. GP permits discovering a game playing strategy (Koza 1992). Figure 1 shows an example of a tree of the computer program performance. This tree corresponds to the program written in the LISP language, as follows: `(LT (* -2.1 x) (LOG y (SQRT x)))`.

This tree is equivalent to the parse tree that most compilers (parsers) construct internally from a computer program source. A parse tree consists of branches and nodes: a root node, a branch node, and a leaf node. A parent node is one which has at least one other node linked by a branch under it. A child node is one which has at least one node directly above it to which it is linked by a branch of the tree.

The size of the parse tree is limited by the number of nodes or by the number of the tree levels. Nodes in the parse tree are divided on functional nodes and terminal ones. A functional node represents the procedure randomly chosen from the primary defined set of functions:

$$\mathcal{F} = \{f_1, \ldots, f_n, \ldots, f_N\} \tag{1}$$

Each function should be able to accept, as its arguments, any value and data type that may possible be returned by the other procedures Koza (1992). Moreover, each procedure should be able to accept any value and data type that may possible be assumed by any terminal in the terminal set:

$$\mathcal{T} = \{a_1, \ldots, a_m, \ldots, a_M\} \tag{2}$$

So, each function should be well defined and closed for any arrangement of arguments that it may come across. Furthermore, the solution to the problem should be expressed by the combination of the procedures from the set of functions and the arguments from the set of terminals. For example, the set of functions $\mathcal{F} = \{AND, NOT\}$ sufficient to express any Boolean function.

3 Selections in Immunological Systems

A biological immune system has elements distributed as well as some features of artificial intelligence like an adaptation, learning, memory, and associative retrieval of information in recognition (Jerne 1984). Especially, the negative selection algorithm (NSA) can be applied for change detection because it uses the discrimination rule to classify some trespassers (Forrest and Perelson 1991). Detectors are randomly generated to reduce those detectors that are not capable of recognizing themselves and detectors capable to distinguish intruders are kept. In the NSA, detection is performed probabilistically (Bernaschi et al. 2006).

An antigen can support an antibody generation and it is a particle that stimulates a reaction against squatters. Besides, some "positive" viruses and bacteria cooperate with antigens (Kim and Bentley 2002). An antigen is recognized by an immunoglobulin (the antibody) that is a huge *Y*-shaped protein capable to recognize and deactivate external objects as "negative" bacteria or virus (Wierzchon 2005). The NSA manages constraints in an evolutionary algorithm by dividing a population in two assemblies (Coello et al. 2002). "Antigens" belong to a feasible solution subpopulation, and "antibodies"—to an infeasible one.

At the beginning of the NSA run an initial fitness for all antibodies in the current infeasible subpopulation is equal to zero. Then, a randomly chosen antigen G^- from the feasible subpopulation is compared to some the selected antibodies. After that, a match measure S between G^- and the antibody B^- is calculated due to a similarity at the genotype level. This measure of the genotype similarity for the chromosome integer coding is, as follows (Balicki 2005):

$$S\left(G^-, B^-\right) = \sum_{m=1}^{M} \left|G_m^- - B_m^-\right|, \tag{3}$$

where

M length of solution,
G_m^- value of the antigen at position m, $m = \overline{1,M}$,
B_m^- value of the antibody at position m, $m = \overline{1,M}$.

The negative selection can be modeled by an adjusted evolutionary algorithm that prefers the infeasible solutions that are similar to a randomly chosen feasible one in the current population. We assume that all random choices of antigens are based on an uniform distribution.

The situation is different in a case of the antibodies. If the fitness of the selected winner is increased by adding the amount of the similarity measure, then the antibody may pass over because of a relatively small value of assessment (3). On the other hand, some constraints may be satisfied by this alternative. What is more, if the constraint is exceeded and the others are not, the value of the similarity measure may be lower for some cases. One of two similar solutions, in genotype sense, may not satisfy this constraint and another may satisfy it.

4 An Improved NSA*

To avoid above mentioned restrictions, some similarity measures are developed from the state of the antibody B$^-$ to the state of the selected antigen G$^-$, as below:

$$f_n(B^-, G^-) = \begin{cases} g_k(B^-) - g_k(G^-), \; k = \overline{1,K}, \; n = k, \\ \left|h_l(B^-)\right|, \; l = \overline{1,L}, \; n = K + l, \end{cases} \quad n = \overline{1,N}, \; N = K + L. \tag{4}$$

where

$$g_k(x) \le 0, \quad k = \overline{1,K},$$
$$h_l(x) = 0, \quad l = \overline{1,L}$$

The distance $f_n(B^-, G^-)$ between B^- and G^- is supposed to be minimized for all constraint indexes n. If $f_n(B^-, G^-) < f_n(C^-, G^-)$, then B^- ought to be preferred to C^- due to the nth constraint. Moreover, if B^- is characterized by the distances to the antigen all shorter than the antibody C^-, then B^- should be selected for all

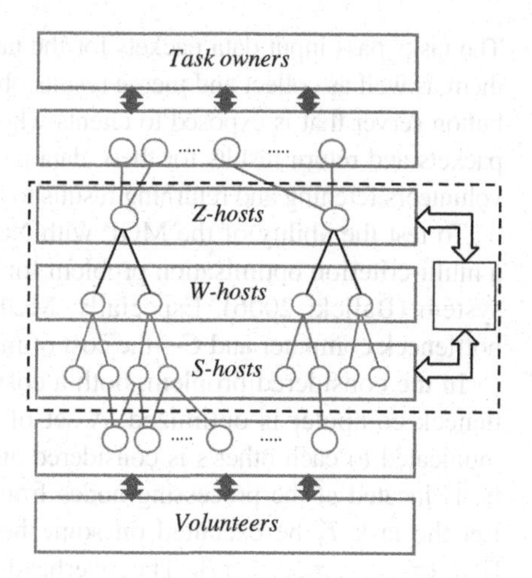

Fig. 2 Architecture of the comcute system

constraints. However, some situations may occur when B^- is characterized by the shorter distances for some constraints and C^- is marked by the shorter distances for the others. In this case, it is difficult to select an antibody. So, a ranking procedure can be applied to calculate fitness of antibodies and then to select a winner.

In the ranking procedure, the distances between the chosen antigen and some antibodies are calculated due to their ranks (Balicki 2006). If B^- is characterized by the rank $r(B^-)$ such that $1 \leq r(B^-) \leq r_{\max}$, then the increment of the fitness function is estimated, as below:

$$\Delta f(B^-) = r_{\max} - r(B^-) + 1 \quad (5)$$

Subsequently, some fitness values of selected antibodies are increased by their given increments. Then antibodies are returned to the current population and this process is repeated typically three times the number of antibodies. Each time, a randomly chosen antigen is compared to the same subset of antibodies.

Afterwards, a new population is constructed by selection, crossover and mutation without calculations of fitness. That process is repeated until a convergence of population emerges or until a maximal number of iterations is exceeded. At the end, the final population as outcomes from the negative selection algorithm is returned to the external evolutionary algorithm.

5 Optimization Model of Volunteer and Grid System

In the architecture of the volunteer and the grid system Comcute (Fig. 2), one can distinguish the *Z*-layer where the system client defines new tasks, starts instances of previously defined tasks, tracks statuses of running tasks and fetches results for completed tasks. On the other hand, *W*-server layer supervises execution of tasks. For each task instance, a subset of *W*-servers is arranged that partitions the task among its members.

The tasks pass input data packets for the task instance to connected *S*-servers beneath them as well as collect and merge results obtained from the *S*-layer. *S*-server is a distribution server that is exposed to clients who fetch execution code and subsequent data packets and return results for these data packets. *I*-client level is an untrusted layer of volunteers fetching and returning results to the system.

To test the ability of the MGP with NSA* for handling constraints, we consider a multi-criterion optimisation problem for task assignment in a distributed computer system (Balicki 2006). Especially, MGP can minimize Z_{max}—a workload of a bottleneck computer and *C*—the cost of machines, concurrently.

In the considered problem, both a cost of hosts as well as a workload of a bottleneck computer is optimized. A set of parallel tasks $\{T_1, \ldots, T_v, \ldots, T_V\}$ communicated to each other's is considered among the coherent computer network with hosts located at the processing nodes from the given set $W = \{w_1, \ldots, w_i, \ldots, w_I\}$. Let the task T_v be executed on some hosts taken from the set of available sorts $\Pi = \{\pi_1, \ldots, \pi_j, \ldots, \pi_J\}$. The overhead performing time of the task T_v by the computer π_j is represented by an item t_{vj}.

The first criterion is a total host's cost, as follows:

$$C(x) = \sum_{i=1}^{I} \sum_{j=1}^{J} \kappa_j x_{ij}^{\pi}, \tag{6}$$

where

$$x = [x_{11}^{m}, \ldots, x_{vi}^{m}, \ldots, x_{VI}^{m}, x_{11}^{\pi}, \ldots, x_{ij}^{\pi}, \ldots, x_{IJ}^{\pi}]^T,$$

$$x_{ij}^{\pi} = \begin{cases} 1 & \text{if } \pi_j \text{ is assigned to the } w_i, \\ 0 & \text{in the other case.} \end{cases}$$

$$x_{vi}^{m} = \begin{cases} 1 & \text{if task } T_v \text{ is assigned to } w_i, \\ 0 & \text{in the other case,} \end{cases}$$

$$\kappa_j = \text{the cost of the host} \pi_j.$$

Another criterion is Z_{max}—a workload of the bottleneck host that is supposed to be minimized. It is provided by the subsequent formula:

$$Z\max(x) = \max_{i \in \overline{1,I}} \left\{ \sum_{j=1}^{J} \sum_{v=1}^{V} t_{vj} x_{vi}^{m} x_{ij}^{\pi} + \sum_{v=1}^{V} \sum_{\substack{u=1 \\ u \neq v}}^{V} \sum_{i=1}^{I} \sum_{\substack{k=1 \\ k \neq i}}^{I} \tau_{vuik} x_{vi}^{m} x_{uk}^{m} \right\}, \tag{7}$$

where τ_{vuik}—total communication time between the task Tv assigned to the ith node and the Tu assigned to the *k*th node.

Figure 3 shows the workload of the bottleneck computer when the function Z_{max} takes value from a period [40; 110] for 256 solutions. What is more, even a small movement of a task to another host or a substitution of the host sort can cause a relatively big alteration of its workload.

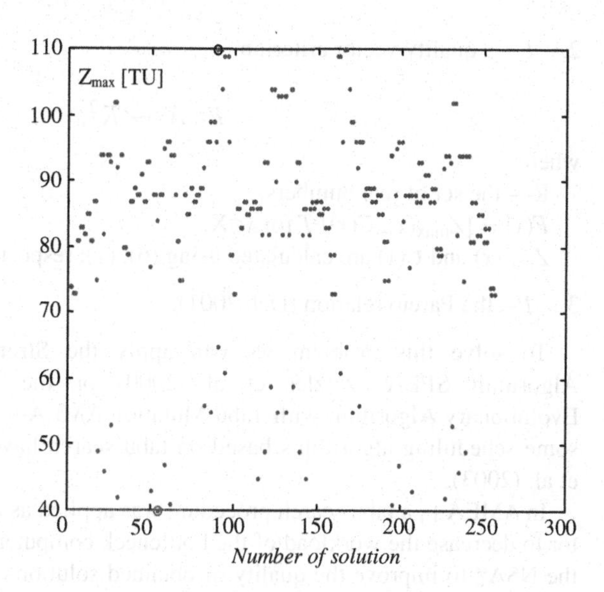

Fig. 3 Workload of the bottleneck computer for generated solutions

The host is supposed to be equipped with necessary capacities of resources. Let the memories $z_1, \ldots, z_r, \ldots, z_R$ be available in the volunteer system and let d_{jr} be the capacity of memory z_r in the host π_j. We assume the task T_v holds c_{vr} units of memory z_r during a program execution. The host memory limit cannot be exceeded in the ith node, as bellows:

$$\sum_{v=1}^{V} c_{vr} x_{vi}^{m} \le \sum_{j=1}^{J} d_{jr} x_{ij}^{\pi}, \quad i = \overline{1,I}, \; r = \overline{1,R}. \tag{8}$$

Let π_j be spoiled independently according to the exponential distribution with rate λ_j. The hosts and the tasks like Z, W or S can be allocated to the nodes to guarantee the required reliability R, as below (Balicki 2005):

$$\prod_{v=1}^{V} \prod_{i=1}^{I} \prod_{j=1}^{J} \exp(-\lambda_j t_{vj} x_{vi}^{m} x_{ij}^{\pi}) \le R_{\min} \tag{9}$$

Let $(\mathcal{X}, F, P)$ be the multi-criterion optimisation question for finding the representation of Pareto-optimal solutions (Coello et al. 2002). It is established, as follows:

1. X—an admissible solution set

$$\mathcal{X} = \{x \in \mathcal{B}^{I(V+J)} | \sum_{v=1}^{V} c_{vr} x_{vi}^{m} \le \sum_{j=1}^{J} d_{jr} x_{ij}^{\pi}, \quad i = \overline{1,I}, \quad r = \overline{1,R};$$

$$\prod_{v=1}^{V} \prod_{i=1}^{I} \prod_{j=1}^{J} \exp(-\lambda_j t_{vj} x_{vi}^{m} x_{ij}^{\pi}) \le R_{\min}; \sum_{i=1}^{I} x_{vi}^{m} = 1, \; v = \overline{1,V}; \sum_{j=1}^{J} x_{ij}^{\pi} = 1, \; i = \overline{1,I}\}$$

where B = {0, 1}.

2. *F*—a quality vector criterion

$$F : \mathcal{X} \to \mathcal{R}^2, \tag{10}$$

where

R— the set of real numbers,

$F(x) = [Z_{max}(x), C(x)]\ T$ for $x \in X$,

$Z_{max}(x)$ and $C(x)$ are calculated using (6), (7), respectively

3. *P*—the Pareto relation (Deb 2001).

To solve this problem we can apply the Strength Pareto Evolutionary Algorithm SPEA (Zitzler et al. 2000) or the Adaptive Multi-Criterion Evolutionary Algorithm with Tabu Mutation AMEA+ (Balicki 2005). Moreover, some scheduling algorithms based on tabu search have been studied in Weglarz et al. (2003).

In AMEA+, a tabu search procedure was applied as the second mutation operator to decrease the workload of the bottleneck computer. Moreover, we introduced the NSA* to improve the quality of obtained solutions and the evolutionary algorithm was denoted as AMEA*.

6 Multi-criterion Genetic Programming

Genetic programming paradigm can be implemented as a genetic algorithm written in the Matlab language. Chromosomes are generated as the Matlab functions and then genetic operators are applied for finding Pareto-suboptimal solutions. Figure 4 shows a scheme of the MGP that operates on the population of program functions. The preliminary population of programs is created in a specific manner (Fig. 4, line 3). Each generated program consists of set of procedures and set of attributes. Set of procedures is defined, as follows:

$$\mathcal{F} = \{list, +, -, *, /\} \tag{11}$$

where *list*—the procedure converting $I(V + J)$ real numbers called *activation levels* to $I(V + J)$ output binary numbers $x^m_{11}, \ldots, x^m_{1I}, \ldots, x^m_{vi}, \ldots, x^m_{VI}, x^\pi_{11}, \ldots, x^\pi_{ij}, \ldots, x^\pi_{IJ}$.

The procedure *list* is an obligatory the root of the program tree and appears only one in a generated program. In that way, the formal constraint $x_m \in \mathcal{B}, m = \overline{1, M}$ is satisfied. An activation level is supplied to a root from the sub-tree that is randomly generated with using arithmetic operators {+, −, *, /} and the set of terminals. Let $\mathcal{D}$ be the set of numbers that consists of the given input data. A terminal set is determined for the problem, as below:

$$\mathcal{T} = \mathcal{D} \cup \mathcal{L}, \tag{12}$$

where $\mathcal{L}$—set of *n* random numbers, $n = \overline{\mathcal{D}}$.

1. BEGIN
2. t:=0, set the even size of the population L, p_m:=1/(ML)
3. generate an initial population of programs $\boldsymbol{P}(t)$
4. run programs, calculate ranks $r(x)$ and fitness $f(x), x \in \boldsymbol{P}(t)$
5. *finish*:=FALSE
6. WHILE NOT *finish* DO
7. BEGIN /* new population */
8. t:= t+1, $\boldsymbol{P}(t) := \varnothing$
9. calculate selection probabilities $p_s(x),\ x \in \boldsymbol{P}(t-1)$
10. FOR L/2 DO
11. BEGIN /* reproduction cycle */
12. *2WT-selection* of a potential parent pair (**a**,**b**) from the population $\boldsymbol{P}(t\text{-}1)$
13. *S-crossover* of a pair (**a**,**b**) with the adaptive crossover rate $p_c := e^{-t/T_{\max}}$
14. *S-mutation* of an offspring pair (**a'**,**b'**) with the mutation rate p_m
15. $\boldsymbol{P}(t)$:=$\boldsymbol{P}(t)\cup$(**a'**,**b'**}
16. END
17. calculate ranks $r(x)$ and fitness $f(x), x \in \boldsymbol{P}(t)$
18. IF ($\boldsymbol{P}(t)$ converges OR $t \geq T_{max}$) THEN *finish*:=TRUE
19. END
20. END

Fig. 4 Multi-criterion genetic programming MGP*

If x is admissible, then the fitness (Fig. 4, line 4) is estimated, as below:

$$f(x) = r_{\max} - r(x) + P_{\max} + 1, \tag{13}$$

where $r(x)$ denotes the rank of an admissible solution, $1 \leq r(x) \leq r_{\max}$.

Moreover, a niching procedure can be applied. The surface region of the Pareto front is divided by the size of the population. The number of other member's falling within the sub-area of any individual is taken to establish the penalty for it.

In the two-weight tournament 2WT selection (Fig. 4, line 12), the roulette rule is carried out twice. If two potential parents (a, b) are admissible, then a dominated one is eliminated. If two solutions non-dominate each other, then they are accepted. If potential parents (a, b) are non-admissible, then an alternative with the smaller penalty is selected.

The quality of solutions increases in optimisation problems with one criterion, if the crossover probability and the mutation rate are changed in an adaptive way (Sheble and Britting 1995). The crossover point is randomly chosen for the chromosome X in the *S-crossover* operator (Fig. 4, line 13). The crossover probability is 1 at the initial population and each pair of potential parents is obligatory taken for the crossover procedure. A crossover operation supports searching high-quality solution areas that it is important in the early search stage. If the number of generation t increases, the crossover probability decreases due to formula $p_c = e^{-t/T_{\max}}$. Some search areas are identified after several crossover operations on parent pairs. That is why, value p_c is smaller

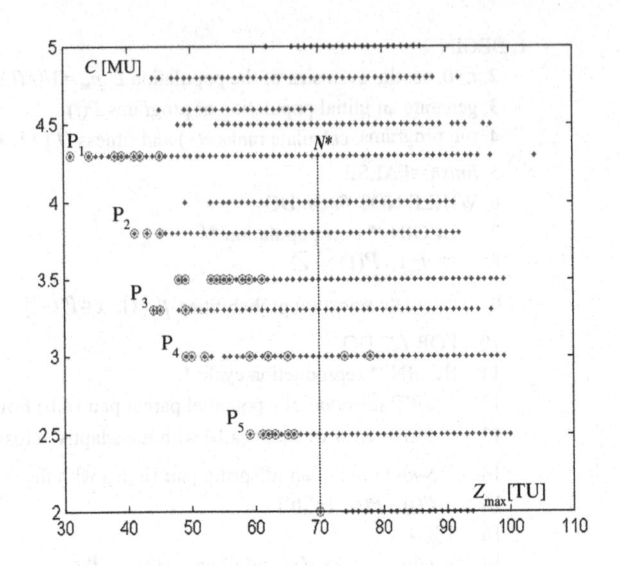

Fig. 5 Pareto front determined by GMP*

and it is equal to 0.6065, if $t = 100$ for maximum number of population $T_{\max} = 200$. The final smallest value p_c is 0.3679. A crossover probability decreases from 1 to *exp*(−1), exponentially. During *S-crossover*, a subtree with the randomly selected root from program *a* is exchanged with another subtree from tree *b*.

In *S-mutation* (Fig. 4, line 14), the random node is chosen as a root, the related subtree is removed, and then a new subtree is generated. A mutation rate is constant in the MGP and it is equal to 1/*M*, where *M* represents the number of decision variables.

7 Numerical Experiments

Better outcomes from the NSA* are transformed into improving a solution quality obtained by the MGP*. For the instance with 15 tasks, 4 nodes, and 5 computer sorts, there are 80 binary decision variables. An average level of convergence to the Pareto set is 17.7 % for the MGP* and 17.4 % for the AMEA*. A maximal level is 28.5 % for the MGP* and 29.6 % for the AMEA*. For this instance the average number of optimal solutions is 19.5 % for the MGP* and 21.1 % for the AMEA*. Figure 5 shows the process of finding efficient task assignment by MGP* for the cut obtained from the evaluation space according to the cost criterion *C* and the workload of the bottleneck computer $Z_{\max}$.

An average level of convergence to the Pareto set, an maximal level, and the average number of optimal solutions become worse, when the number of task, number of nodes, and number of computer types increase. An average level is 37.7 % for the MGP* versus 35,7 % for the AMEA*, if the instance includes 50 tasks, 4 nodes, 5 computer types and also 220 binary decision variables.

8 Concluding Remarks

Multi-objective genetic programming is a relatively new paradigm of artificial intelligence that can be used for finding Pareto-optimal solutions. A computer program as a chromosome gives possibility to represent knowledge that is specific to the problem in more intelligent way than the data structure.

Our future works will focus on testing other sets of procedures and terminals to find the Pareto-optimal task assignments for different criteria and constraints. Initial numerical experiments confirmed that sub-optimal in Pareto sense task assignments can be found by genetic programming. This approach permits obtaining outcomes of comparable quality to advanced evolutionary algorithm.

References

Balicki J (2005) Immune systems in multi-criterion evolutionary algorithm for task assignments in distributed computer system. Lect Notes Comput Sci 3528:51–56

Balicki J (2006) Multicriterion genetic programming for trajectory planning of underwater vehicle. J Comput Sci Netw Secur 6:1–6

Bernaschi M, Castiglione F, Succi S (2006) A high performance simulator of the immune system. Future Gener Comput Syst 15:333–342

BOINC. Open-source software for volunteer and grid computing. http://boinc.berkeley.edu/. Accessed 25 Oct 2013

Coello CAC, Van Veldhuizen DA, Lamont GB (2002) Evolutionary algorithms for solving multi-objective problems. Kluwer Academic Publishers, New York

Deb K (2001) Multi-objective optimization using evolutionary algorithms. Wiley, Chichester

Forrest S, Perelson AS (1991) Genetic algorithms and the immune system. Lect Notes Comput Sci 496:319–325

Jerne NK (1984) Idiotypic networks and other preconceived ideas. Immunol Revue 79:5–25

Kim J, Bentley PJ (2002) Immune memory in the dynamic clonal selection algorithm. In: Proceedings of 1st international conference on artificial immune systems, Canterbury, Australia, pp 57–65

Koza JR (1992) Genetic programming: on the programming of computers by means of natural selection. MIT Press, Cambridge

Koza JR, Keane MA, Streeter MJ, Mydlowec W, Yu J, Lanza G (2003) Genetic programming IV. Routine human-competitive machine intelligence. Kluwer Academic Publishers, New York

Samuel AL (1960) Programming computers to play games. Adv Comput 1:165–192

Sheble GB, Britting K (1995) Refined genetic algorithm—economic dispatch example. IEEE Trans Power Syst 10:117–124

Weglarz J, Nabrzyski J, Schopf J (2003) Grid resource management: state of the art and future trends. Kluwer Academic Publishers, Boston

Wierzchon ST (2005) Immune-based recommender system. In: Hryniewicz O, Kacprzyk J, Koronacki J, Wierzchon ST (eds) Issues in intelligent systems. Paradigms. Exit, Warsaw, pp 341–356

Zitzler E, Deb K, Thiele L (2000) Comparison of multiobjective evolutionary algorithms: empirical results. Evol Comput 8:173–195

Particle Swarm Optimization Techniques: Benchmarking and Optimal Power Flow Applications

C. Barbulescu and S. Kilyeni

Abstract Nowadays, there is a great interest in using of artificial intelligence techniques in different areas. The current work is focusing on particle swarm optimization (PSO) study. The authors aim to present a synthesis regarding the PSO applications within the power system field. Two issues are addressed in the chapter. First, the PSO parameter tuning procedure using mathematical test functions is presented. Second, the conclusions are elaborated in a case of optimal power flow (OPF) computing for large scale test power systems. For both issues the methodologies and software tools developed are presented. The research work is continued focusing on development the PSO based software designed for transmission network expansion (in case of complex power systems).

1 Introduction

Nature represents the main background for swarm type intelligence methods: fish tanks, flocks of birds or ant colonies. All these examples are proving extraordinary self-organizing capabilities. The collective behavior is not able to be described adequately only by gathering each member's individual one.

Swarm type intelligence appeared in 1989 within the optimization field as a set of autonomous robots control algorithms (Beni and Wang 1993). Six years later, three main algorithms have been developed: ant colony optimization (ACO), stochastic diffusion search (SDS) and particle swarm optimization (PSO). The genetic algorithms and the swarm intelligence based ones form the evolutionary computation domain.

C. Barbulescu (✉) · S. Kilyeni
Department of Power Systems, Politehnica University Timisoara, Timisoara, Romania
e-mail: constantin.barbulescu@upt.ro

S. Kilyeni
e-mail: stefan.kilyeni@upt.ro

Z. S. Hippe et al. (eds.), *Issues and Challenges in Artificial Intelligence*, Studies in Computational Intelligence 559, DOI: 10.1007/978-3-319-06883-1_12,

Swarm type intelligence describes collective systems, formed by simple agents, and characterized by a specific intelligence. This system is able to self-organize, based on the agent-agent type local interactions and, also, agent-environment ones. The population is called *swarm* and its individuals are called *particles*.

PSO has been developed by Kennedy and Eberhart (1995). The first application of the PSO has been used for the artificial neural network training process (Kennedy and Eberhart 1995). The corresponding algorithm has been developed and, further, it is also utilized in a telecommunication, data mining, designing, optimization, power engineering, signal processing etc. It has an advantage of being simple in concept, easy to implement and computationally efficient (AlRashidi and El-Hawary 2007; Esmin et al. 2005; Yumbla et al. 2008). In Zhao et al. (2004) PSO is applied for solving the OPF however, only using a continuous control approach. Meanwhile, in He et al. (2004) discrete variables are also included by adding penalty terms to an objective function.

The chapter is organized as follows. The Sect. 2 introduces and describes the PSO mechanism. The developed software tool is presented in the Sect. 3. The Sect. 4 presents the results and their discussion. Conclusions are presented in the Sect. 5.

2 PSO Mechanism

Two research directions have been studied and are presented in the paper: evaluation of the mathematical test functions using PSO-based algorithm and OPF computation based on the PSO.

2.1 Standard PSO

The algorithm for the standard PSO is presented. The following function is considered:

$$f(x) : X \rightarrow Y \subseteq R^n \tag{1}$$

where: $x = \{x_1, x_2, x_3, \ldots, x_d\}$—vector containing d variables. The goal is to compute the solution that provides minimum of the function. The swarm is defined as:

$$\mathbf{S} = \{x_1, x_2, \ldots, x_{np}\} \tag{2}$$

where: np—particle number (each particle being a possible solution candidate). It is an important parameter, representing the swarm dimension. Each particle is described by a set of d variables:

$$x_i = \{x_{i,1}, x_{i,2}, \ldots, x_{i,d}\} \in A, \quad i = 1, 2, \ldots, np \tag{3}$$

where: A—set of the possible solutions.

To find the solution, each particle is going to move iteratively within the searching space, $\boldsymbol{x}_i$ representing the particle position. The position is updated using the v velocity value:

$$x_i = \{x_{i,1}, x_{i,2}, \ldots, x_{i,d}\} \in A, \quad i = 1, 2, \ldots, np \tag{4}$$

The velocity components are computed for each variable. For a specific computing step t each particle is characterized by the x_i^t position and the v_i^t velocity.

The swarm uses two memories. The first memory refers to each particle and it is known as *personal best* (*pBest*).

$$f(\mathbf{pBest}_i^{t+1}) = Min\{f(x_i^1), f(x_i^2), \ldots, f(x_i^t)\}, \ i = \overline{1, np} \tag{5}$$

$$\mathbf{pBest}_i^{t+1} = \{pBest_{i,1}^{t+1}, pBest_{i,2}^{t+1}, \ldots, pBest_{i,d}^{t+1}\}, \ i = \overline{1, np} \tag{6}$$

where: t—current computing step.

The PSO algorithm social behaviour is described by the second memory called *global best* (*gBest*).

$$f(g\mathbf{Best}^{t+1}) = Min\{f(\mathbf{pBest}_1^{t+1}), f(\mathbf{pBest}_2^{t+1}), \ldots, f(\mathbf{pBest}_{np}^{t+1})\} \tag{7}$$

$$\mathbf{gBest}^{t+1} = \{gBest_1^{t+1}, gBest_2^{t+1}, \ldots, gBest_d^{t+1}\} \tag{8}$$

One PSO algorithm computing step is described as:

$$v_{i,j}^{t+1} = v_{i,j}^t + c_1 \cdot r_1 \cdot (pBest_{i,j}^t - x_{i,j}^t) + c_2 \cdot r_2 \cdot (gBest_j^t - x_{i,j}^t), i = \overline{1, np}, j = \overline{1, d} \tag{9}$$

$$x_{i,j}^{t+1} = v_{i,j}^t + x_{i,j}^{t+1}, \quad i = 1, 2, \ldots, np, j = 1, 2, \ldots, d \tag{10}$$

where: r_1 and r_2—random variables uniformly distributed within the [0; 1] range; c_1 and c_2—acceleration constants corresponding to the cognitive and social terms.

The values corresponding to the c_1 and c_2 constants have an important influence over the particles' behaviour within the searching space (usual values are set within [1; 2] range). Few practical considerations are the following ones:

- close values to *2* determine the particles to explore farther areas within the searching space;
- in case of $c_1 > c_2$ the searching is forced towards the particles' personal best values (*pBest*);
- in case of $c_1 < c_2$ the searching mechanism is oriented towards the *gBest* value. It is ideal for convex functions (without local minimum; only global one).

The acceleration constants' values are established based on experience. To improve the PSO performance, additional searching space or velocity constraints can be set:

- *infinite searching*: particle are allowed to evolve towards an unfavourable area. Thus a valid solution is not able to be established. The velocity and position are not changing;
- *absorption searching*: particles that are situated outside the admissible domain are reinitialized with the nearest position to the domain border;
- *random searching*: particles that are situated outside the admissible domain are randomly reinitialized inside the domain. In the following, the particles' velocity is adjusted according to the relation:

$$v_{i,j}^{t+1} = x_{i,j}^{t+1} - x_{i,j}^{t} \tag{11}$$

For the great majority of the optimization applications it is recommended that the particles should not exceed the admissible domain. Thus, the absorption and random searching method are suggested.

Very high velocity values represent one of the problems that could arise in case of PSO algorithm evolution. In this case the particles are tending to evolve outside the admissible range. Velocity limiting represents a solution for solving this problem.

The admissible domain is established as follows:

$$x_i \in [a_i, b_i] \tag{12}$$

where a_i and b_i represent the inferior and superior limits for x_i variable. The following velocity limits are obtained:

$$v_j^{max} = \frac{b_j - a_j}{k}, \quad j = \overline{1,n} \; v_j^{min} = -v_j^{max}, \quad j = \overline{1,n} \tag{13}$$

The k coefficient value is established according to the experience. The most recommended value is $k = 2$, leading to an efficient searching space exploring. In case of a non convex function (that has multiple local minimum values) greater k coefficient values are recommended to be used.

Once the velocities have been computed using relation (9), the obtained values are checked and the necessary corrections are performed:

$$v_{i,j}^{t+1} = \begin{cases} v_{i,j}^{t+1} & \text{if } v_{i,j}^{t+1} \in [v_j^{min}, v_j^{max}] \\ v_j^{min} & \text{if } v_{i,j}^{t+1} < v_j^{min}, \\ v_j^{max} & \text{if } v_{i,j}^{t+1} > v_j^{max} \end{cases}, \quad i = \overline{1,n_p}, \quad j = \overline{1,d} \tag{14}$$

2.2 PSO Variants

In case of PSO algorithm, like other population based searching procedures, two major stages are highlighted: *exploration*—the most suitable areas within the searching space are identified and *exploitation*—convergence through the best solution is assured.

Two PSO variants are divided: *global PSO*—entire swarm is considered as neighbours of each particle and a single global value characterizing the best position of the swarm is used—and *local PSO*—characterized by several positions indicated by the neighbouring groups formed within the swarm. Usually, global PSO provides an accelerated convergence towards the optimal value. Local PSO performs a more detailed search within the solution space, thus better exploration is assured.

In case global PSO, for a swarm having n_p particles and n variables solution space, the x_i particle velocity is computed as:

$$vg_{i,j}^{t+1} = v_{i,j}^{t} + c_1 \cdot r_1 \cdot (pBest_{i,j}^{t} - x_{i,j}^{t}) + c_2 \cdot r_2 \cdot (gBest_{j}^{t} - x_{i,j}^{t}),\ i = \overline{1, n_p}, j = \overline{1, d} \tag{15}$$

and for *local PSO* variant the corresponding relation is:

$$vl_{i,j}^{t+1} = v_{ij}^{t} + c_1 \cdot r_1 \cdot (pBest_{i,j}^{t} - x_{i,j}^{t}) + c_2 \cdot r_2 \cdot (iBest_{j}^{t} - x_{i,j}^{t}),\ i = \overline{1, n_p}, j = \overline{1, d} \tag{16}$$

Using this two information a new PSO algorithm variant is developed—the *unified PSO* (UPSO). The two previous velocities are combined through the unification factor u:

$$v_{i,j}^{t+1} = u \cdot vg_{i,j}^{t+1} + (u - 1) \cdot vl_{i,j}^{t+1},\ i = \overline{1, n_p},\ j = \overline{1, d} \tag{17}$$

where $u \in [0, 1]$.

For small u values the searching space is explored more efficiently. If a value close to 1 is used, then the searching space is smaller, the convergence being accelerated (global PSO velocity influence is transmitted) (Bijaya et al. 2001).

2.3 PSO Based OPF Computing

The control variables are represented by:

$$V_i, i \in G, P_{gi}, i \in G \backslash e, k_{ij}, ij \in T \tag{18}$$

where: V_i—*i*th PV bus voltage magnitude; P_{gi}—*i*th bus real generated power; k_{ij}—transformer ratio for the ij transformer; G—subset of the PV buses; e—the slack bus; T—subset of the transformers.

The state variables are represented by:

$$\delta_i, i \in N \backslash e, P_{ge}, V_i, i \in C, Q_{gi}, i \in G \tag{19}$$

where: δ_i—ith bus voltage angle; V_i—ith PQ bus voltage magnitude; Q_{gi}—ith bus reactive generated power; N—set of buses; C—subset of PQ buses.

Constraints are represented by:

- equality constraints:

$$P_i(\mathbf{U}, \boldsymbol{\delta}, \mathbf{K}) - P_{gi} - P_{ci} = 0, i \in N; Q_i(\mathbf{U}, \boldsymbol{\delta}, \mathbf{K}) - Q_{gi} - Q_{ci} = 0, \quad i \in N \tag{20}$$

where: $\boldsymbol{V}$, $\boldsymbol{\delta}$, $\boldsymbol{K}$—bus voltage, bus voltage angle and transformer ratio arrays.

- inequality constraints:

$$x^{\min} \leq x \leq x^{\max} \tag{21}$$

where: variable x, respectively refers to: P_{ge}, Q_g, V, P_{ij}, S_{ij}, P_g, k.

The objective function refers to:

$$\min(OBF) = \min\left(\sum_{i \in G} P_{gi} + \sum_{i \in N} P_{ci}\right) \tag{22}$$

In this case the real power losses are subject to minimization. This problem is solved by using penalty functions and Lagrange multipliers method. For this purpose, the lagrangean function Φ is build:

$$\begin{aligned}
\Phi =& (\sum_{i \in G} P_{gi} + \sum_{i \in G} P_{ci}) + \sum_{i \in N \backslash e} \lambda_{pi} \cdot (P_i - P_{gi} - P_{ci}) + \sum_{i \in C} \lambda_{qi} \cdot (Q_i - Q_{ci}) + \\
&+ r_{pe} \cdot (P_{ge} - P_{ge}^*)^2 + r_q \cdot \sum_{i \in G} p_{qi} \cdot (Q_{gi} - Q_{gi}^*)^2 + r_V \cdot \sum_{i \in C} p_{Vi} \cdot (V_i - V_i^*)^2 + \\
&+ r_p \cdot \sum_{ij \in R} p_{pij} \cdot (P_{ij} - P_{ij}^*)^2 + r_s \cdot \sum_{ij \in R} p_{sij} \cdot (S_{ij} - S_{ij}^*)^2
\end{aligned} \tag{23}$$

where: λ—Lagrange multipliers; r—penalty coefficients; p—weighting coefficients.

The OPF mathematical model is solved basing on PSO approach. The particle is represented by an array having as components the control variables: PV bus voltage magnitudes (V_i, $i \in G$), PV bus real generated power, excepting the slab bus (P_{gi}, $i \in G \backslash e$), transformer ratio values and phases (K_{ij}, $ij \in T$, Ω_{ij}, $ij \in T$). The x_i particle and the swarm S are defined as:

$$S = [x_1, x_2, \ldots, x_N] \tag{24}$$

$$\begin{aligned} x_i &= [x_{i1}, x_{i2}, \ldots, x_{id}]^T, \; i = \overline{1, N}; \\ d &= g + (g - 1) + 2c \end{aligned} \tag{25}$$

$$\begin{aligned} x_i = & [\{U_{i1}, U_{i2}, \ldots, U_{ij}\}, \{Pg_{i1}, Pg_{i2}, \ldots, Pg_{ik}\}, \{K_{i1}, K_{i2}, \ldots, K_{im}\}, \\ & \{\Omega_{i1}, \Omega_{i2}, \ldots, \Omega_{im}\}], \quad i = \overline{1, N}; \quad j \in G; \quad k \in G \backslash e; \; m \in T \end{aligned} \tag{26}$$

where N—swarm dimension; d—number of variables included within the ith particle; c—number of the transformers within the power system. The swarm dimension should not be smaller than 30 particles or, in case of large power systems, at least equal to number d.

Particles are evaluated based on the relation (23) having as an objective the Lagrangean function minimization. Two stopping criteria are considered. The final solution is obtained when the maximum number of iterations is reached or no further improvement can be brought.

In the following, the OPF algorithm is applied. Based on the algorithm a corresponding software tool has been developed.

3 Software Tool

3.1 *Particle Swarm Optimization-benchmark (PSO-b)*

The software tool *PSO-b* has been developed in Matlab (Fig. 1).

Three mathematical functions (Rosenbrock, Rastrigin, Schwefel) are implemented. The user is able to view the plot for the desired function, to change the default domain and to chose the desired number of variables (*varNo* field). The basic PSO parameters are set using the PSO Parameters zone:

- particle numbers composing the swarm—*Swarm Size* field;
- c_1 constant for the cognitive velocity term—*C1(cognitive)* field;
- c_2 constant for the social velocity term—*C2(social)* field.

Additionally, other options are also provided by the developed software tool:

- Search Space Control (random, absorb, infinity options);
- Velocity Control (adaptive velocity and velocity limitation);
- algorithm type is suitable to be chosen: standard PSO or unified PSO;
- in case of unified PSO, 3 configuration types for the unification factor are able to be selected: constant variation, linear variation and exponential variation;
- Number of cycles—field allowing the number of consecutive runs;
- if the Real time graph option is checked, the user is able to view in real time (for each computing step) the particle position within the searching space.

3.2 *PSO Based OPF Computing Software Tool*

The main window is presented in Fig. 2.

The user is able to select the control variables (P, V, k or different combinations) for the optimization process, the swarm size and to set the stopping criteria. The software is linked to other power systems analysis software (such as Power world) to extract the associated database. The stopping criteria refer to the maximum iterations and capping iterations (the OBF value is not improved anymore).

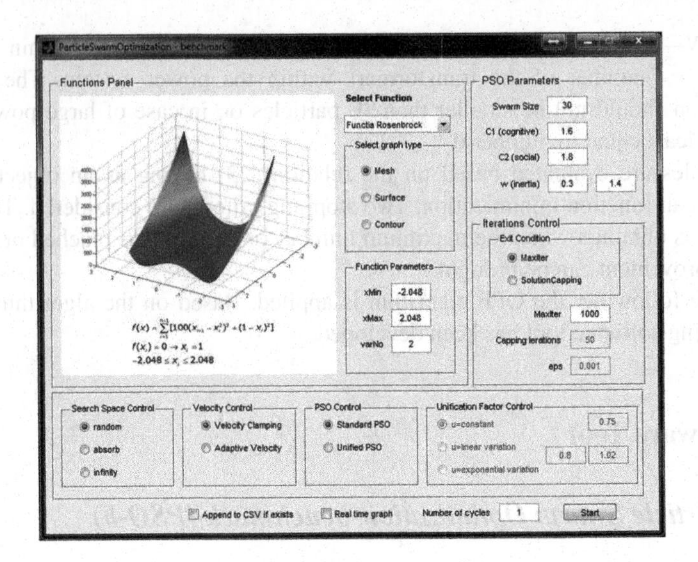

Fig. 1 PSO-b—main window

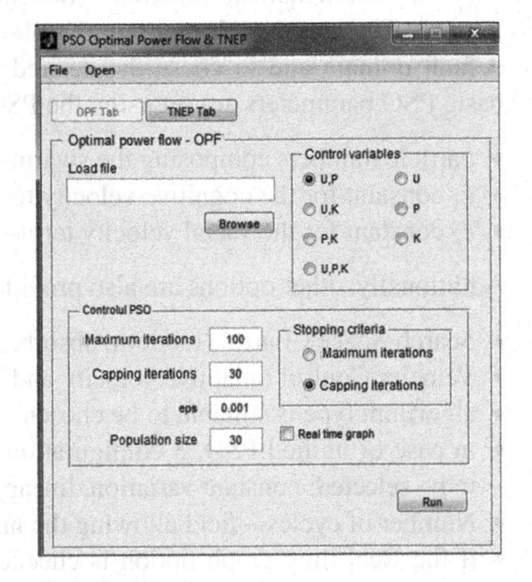

Fig. 2 PSO based OPF software tool

4 Results and Discussions

4.1 Mathematical Test Functions: Rosenbrock Function

Rosenbrock valley represents a classical optimization problem. The global optimum is situated inside a long and narrow valley. The searching algorithms reach the valley easily, but convergence to the global optimum is difficult. This function is used for the searching algorithm performance testing (Fig. 3).

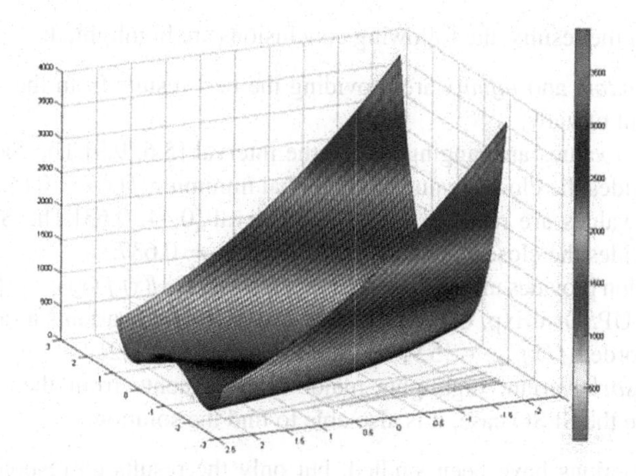

Fig. 3 Rosenbrock's function plot—2 variables

Table 1 SPSO (20 variables, 50 samples)—Rosenbrock functionRosenbrock function

	SPSOr	SPSOa	SPSOi	SaPSOr	SaPSOa	SaPSOi
Best *f(x)* value	0.124	0.000	1.298	0.304	0.000	2.189
Average *f(x)* value	5.872	59.363	8.760	5.677	42.634	9.955
Average *x* value	0.637	0.565	0.502	0.626	0.673	0.441
Average computing time (CT) (s)	08	06	05	08	06	05

Table 2 UPSO (20 variables, 50 samples)—Rosenbrock function

	UPSOr-uCt	UPSOa-uCt	UaPSOr-uCt	UaPSOa-uCt	UPSOr-uExp	UPSOa-uExp	UaPSOr-uExp	UaPSOa-uExp
Best *f(x)* value	1.162	1.944	0.236	0.026	0.032	0.000	3.474	0.000
Average *f(x)* value	7.508	11.970	7.702	9.012	6.775	11.960	5.996	20.224
Average *x* value	0.551	0.585	0.553	0.517	0.595	0.588	0.624	0.588
Average CT (s)	10	8	10	8	10	8	10	8

$$f(x_i) = \sum_{i=1}^{n-1} [100 \cdot (x_{i+1} - x_i^2)^2 + (1 - x_i)^2] \ , \quad -2.048 \leq x_i \leq 2.048 \quad (27)$$

$$f(x_i) = 0 \, , \; x_i = 1 \, , \; i = \overline{1, n} \quad (28)$$

Global minimum is located for $x_m = 1$, with $f(x_m) = 0$ (Tables 1 and 2).

Analyzing the results, the following conclusions are highlighted:

- options *random* and *infinity* are providing the best results from the average *f(x)* values point of view;
- average *f(x)* values are ranging within the interval [5.6, 9.9].The SaPSOr algorithm provides the closest value to the global minimum $f(x_i) = 0\,(5.677)$;
- average x values are ranging within the interval [0.44, 0.63].The SPSOr algorithm provides the closest value to the solution $x_i = 0.637$;
- *absorb* option provides unsatisfied results for the average *f(x)* $f(x)_{avg} \in [42.6, 59.3]$;
- in case of UPSOr-uExp, UPSOr-uCt and UaPSOr-uExp variants a range restriction is recorded $f(x)_{med} \in [5.9, 7.5]$ and $x_{med} \in [0.55, 0.62]$;
- UPSO *absorb* variant highlights major improvements from the average *f(x)* value. Like the SPSO case, it is also able to find the solution.

All the 3 functions have been studied, but only the results corresponding to the Rosenbrock function are presented in the paper. The use of *absorb* option, in case of Rosenbrock function, leads to the global minimum (*0* being the best result). But, the average value is much higher than the one corresponding to *random* option. Thus, for this function, it is recommended to control the search space using *random* option, because the best results are provided and their variation is much smaller in comparison with *absorb* option.

It could be concluded that the best results are provided for the *unified* version of the standard PSO *algorithm*, having a *random* control of the search space, adaptive velocity and an exponential variation unification factor.

4.2 PSO Based OPF Computing

A real large scale power system has been used as case study. It is designed based on the Western and Southern side of the Romanian Power System (Fig. 4). It has 88 buses and 107 branches. The 35 PV buses are divided in 17 real generating units and 18 equivalent PV buses, obtained by extracting the analyzed part from the Romanian Power System. The system has 42 PQ buses. The buses at medium voltage (real generating groups), 110, 220, 400 kV are represented.

A comparison of the bus voltages (base case and OPF) is presented in Fig. 5. None of the voltage limits are exceeded. It is highlighted that the OPF voltage system buses has greater values than the base case.

The real/reactive generated power OPF results are synthesized in Figs. 6 and 7. Analyzing the figures, it is highlighted that the superior or inferior limits are not violated for any of the generating units. Regarding the power flow through the system branches, there are no branches loaded at limit or congestions.

PSO based OPF algorithm settings have been established as follows: population = 30 particles; maximum allowable computing steps = 1,000, executed computing steps = 721; imposed error = 1E−03.

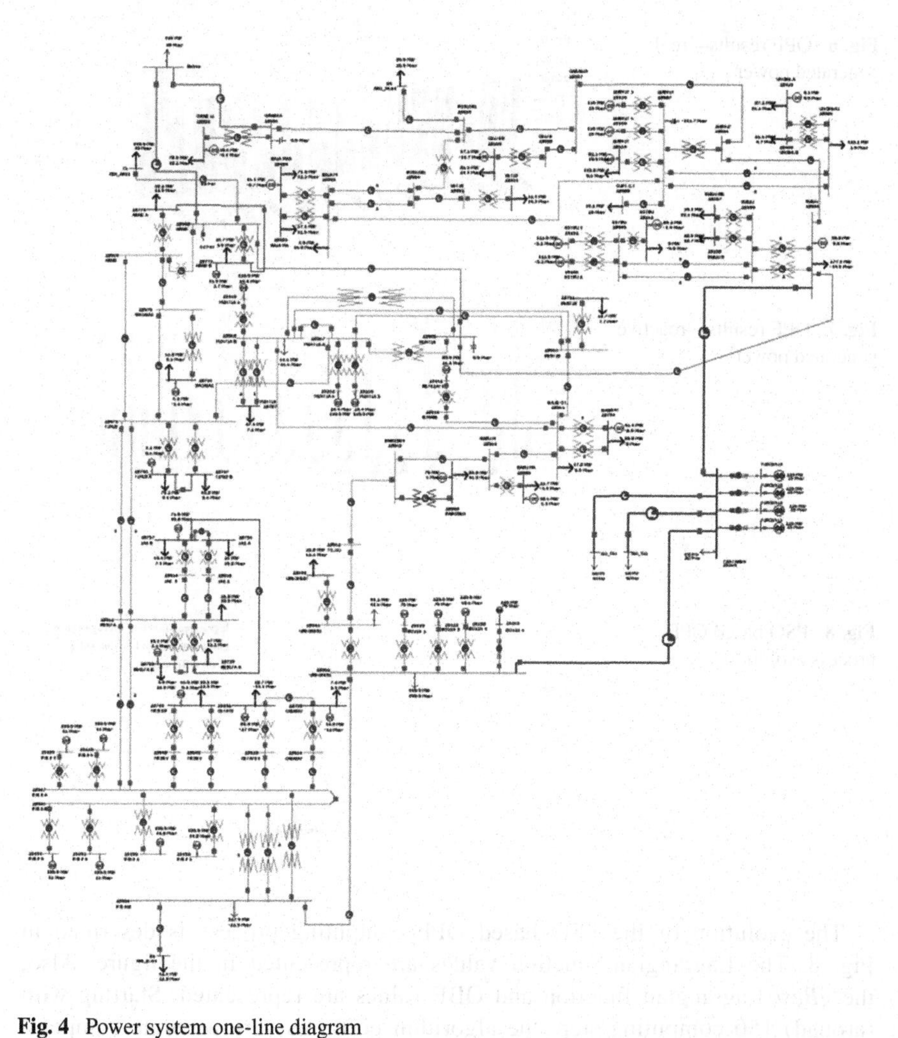

Fig. 4 Power system one-line diagram

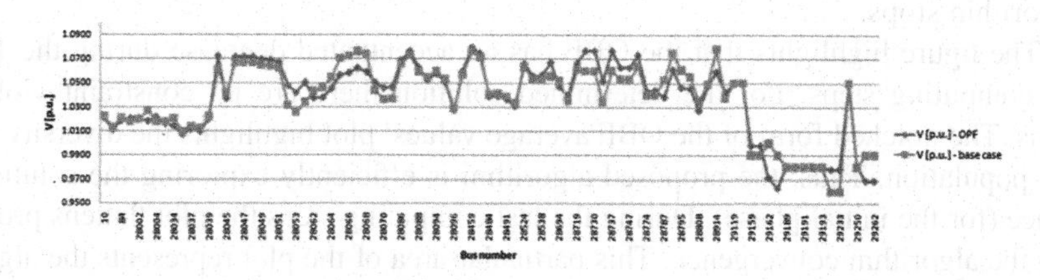

Fig. 5 OPF results—power system buses

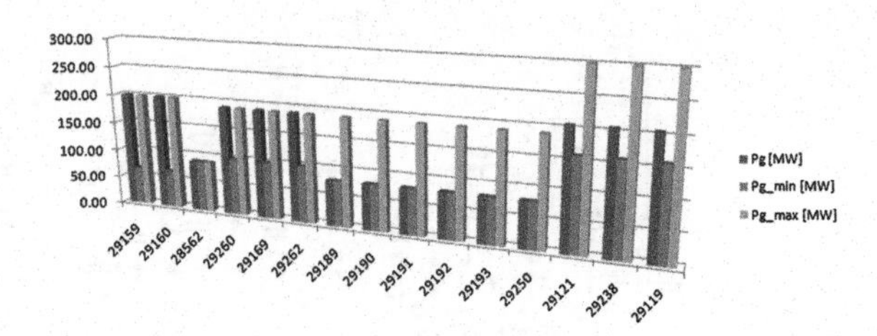

Fig. 6 OPF results—real generated power

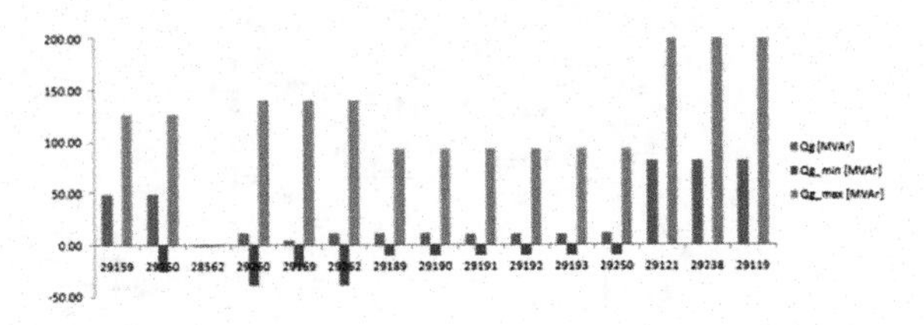

Fig. 7 OPF results—reactive generated power

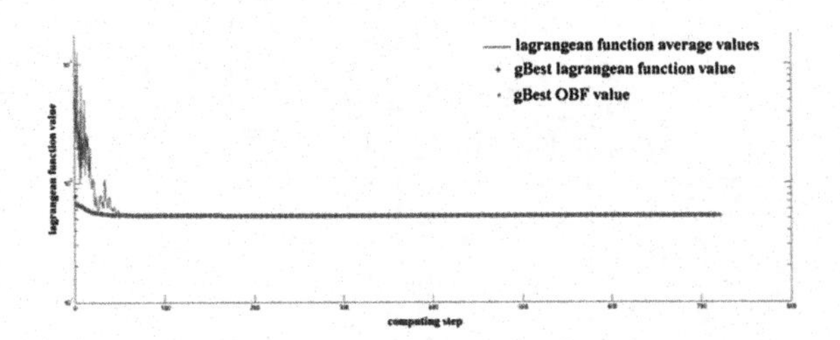

Fig. 8 PSO based OPF process evolution

The evolution of the PSO based OPF computing process is described in Fig. 8. The Lagrangian function values are represented in the figure. Also, the *gBest* Lagrangian function and OBF values are represented. Starting with (around) 150 computing step, the algorithm performs very accurate computations. The solution is not improving only for the last computing steps and the algorithm stops.

The figure highlights that the OBF has an accentuated decrease during the 1st 20 computing steps. For the determined solution there are no constraint violations. The notched form of the OBF average values' plot highlights the diversity of the population. Thus, the proposed algorithm is efficiently exploring the solution space (for the initial phase). During the last computing steps the plot flattens proving the algorithm convergence. This particular area of the plot represents the algorithm computing effort trying to improve the solution having as a goal to minimize the constraints' violations (OBF $\cong$ Lagrangian function).

5 Conclusion

For of PSO based mathematical test functions' evaluation the best results are provided for the unified version of the standard PSO algorithm, having a random control of the search space, adaptive velocity and an exponential variation unification factor.

A real and large scale test of power system has been led as the case study. Useful results have been obtained. The algorithm is characterized by good properties.

For PSO based algorithms, the potential solutions (particles) are performed using a random velocity, spreading through the problem space by following the best particles (having as a goal to improve them). The particles, having random velocities, are able to evolve towards the global optimum basing on their memory mechanism.

References

AlRashidi MR, El-Hawary ME (2007) Hybrid particle swarm optimization approach for solving the discrete OPF problem considering the valve loading effects. IEEE Trans on Power Syst 22(4):2030–2038

Beni G, Wang J (1993) Swarm intelligence in cellular robotic systems, NATO ASI Series, Series F: Computer and System Science 102: 703–712

Bijaya KP, Yuhui S, Meng-Hiot L (2001) Handbook of swarm intelligence, concepts principles and applications. Springer, Berlin

Esmin AAA, Lambert-Torres G, de Souza Zambroni AC (2005) A hybrid particle swarm optimization applied to loss power minimization. IEEE Trans on Power Syst 20(2):859–8662

He S, Wen JY, Prempain E, Wu QH, Fitch J, Mann S (2004) An improved particle swarm optimization for optimal power flow. In: Proceedings of the inter conference on power systems technology, 2: 1633–1637

Kennedy J, Eberhart RC (1995) Particle swarm optimization. In: Proceedings of the IEEE international conference on neural networks, pp 1942–1948

Onate Yumbla PE, Ramirez JM, Coello Coello CA (2008) Optimal power flow subject to security constraints solved with a particle swarm optimizer. IEEE Trans on Power Systems 23(1):33–40

Zhao B, Guo CX, Cao YJ (2004) Improved particle swarm optimization algorithm for OPF problems. In: Proceedings of the IEEE/PES power systems conference exposition, pp 233–238

Time Requirements of Optimization of a Genetic Algorithm for Road Traffic Network Division Using a Distributed Genetic Algorithm

T. Potuzak

Abstract This paper describes the optimization of a dividing genetic algorithm (DGA). It is used for division of road traffic networks into sub-networks of a distributed road traffic simulation. The optimization is performed by finding optimal settings of the DGA parameters using a distributed optimizing genetic algorithm (distributed OGA). Since the distributed OGA is expected to be extremely time-consuming, the paper is focused on a determination of the total time necessary for the OGA computation. It is determined, performing tests, that the OGA can be completed in range of days at least for lower numbers of OGA generations on a distributed computer consisting of nearly 100 processor cores.

1 Introduction

The computer road traffic simulation is an important tool for analysis, control, and design of road traffic networks. Such simulation can be very time consuming, especially for large road traffic networks (e.g. large cities or entire states). Therefore, many road traffic simulators have been adapted or designed for distributed computing environment where combined computing power of multiple interconnected computers (nodes) is used for faster execution of the simulation. In this case, the simulated road traffic network must be divided into sub-networks, whose simulations are then performed on particular nodes of the distributed computer as processes. The quality of the division influences the resulting performance of the entire distributed simulation. Hence, two important issues should be considered during the division—the sub-networks load-balancing and the inter-process communication minimization (Potuzak 2011).

T. Potuzak (✉)
Department of Computer Science and Engineering, Faculty of Applied Sciences,
University of West Bohemia, Plzen, Czech Republic
e-mail: tpotuzak@kiv.zcu.cz

Z. S. Hippe et al. (eds.), *Issues and Challenges in Artificial Intelligence*,
Studies in Computational Intelligence 559, DOI: 10.1007/978-3-319-06883-1_13,

During our previous research, we have developed a method for road traffic network division, which considers both mentioned issues (Potuzak 2011). This method utilizes a genetic algorithm for multi-objective optimization of the resulting road traffic network division. Although the results of the proposed method are reasonable (Potuzak 2012a) they are not optimal, especially for large road traffic networks (Potuzak 2013a). The probable cause is the suboptimal settings of the parameters of the genetic algorithm that are set manually based on preliminary testing result. These settings can be tuned in order to achieve better performance of the division method.

In this paper, we discuss the time requirements of a distributed genetic algorithm for optimization of the parameters settings of the genetic algorithm for road traffic network division. To avoid confusion in the following text the genetic algorithm of optimization is referred as optimizing genetic algorithm (OGA). The genetic algorithm for road traffic network division (which is optimized using OGA) is referred as dividing genetic algorithm (DGA).

2 Basic Notions

In order to make further reading more clear the genetic algorithms and the road traffic network division are described first.

2.1 *Genetic Algorithms Description*

A genetic algorithm is an evolutionary algorithm (Poli et al. 2008), which mimics natural genetic evolution and selection in nature in order to solve a specific problem. Since their development in 1975, the genetic algorithms have been widely used for solving of optimization (including multi-objective optimization) and searching problems in many domains (Farshbaf and Feizi-Darakhshi 2009). The basic notions of a typical genetic algorithm are described in following paragraphs.

Using the genetic algorithm for solving of a problem the representation of the solution of the problem must be determined first. Each solution is considered an *individual* and it is usually represented by a vector of binary or integer values. The meaning of the particular values depends on the solved problem. A predetermined number of individuals are then (most often) randomly generated in order to form the *initial population* (Menouar 2010).

For each individual of the initial population, the so-called *fitness value* is calculated using a *fitness function*. The fitness value is an objective assessment of the quality of each individual from the point of view of the solved problem. The better the solution is, the higher its fitness value should be. So, the fitness function is the only part of the genetic algorithm requiring the knowledge of the solved problem (Menouar 2010). Based on the fitness value, a set of individuals are selected to be "parents" of a new generation.

The new generation is created using the selected parents and the *crossover* and *mutation* operators. The crossover uses (most often) two parents to produce (most often) two descendants using mutual exchange of the values of the parents' vectors. The mutation performs random changes of the values in the descendants' vectors (Poli et al. 2008). Once the new generation is created, the fitness values are calculated for all individuals and the process repeats until a stop condition is fulfilled or a preset number of generations is created (Potuzak 2011).

2.2 Road Traffic Network Division Description

As it is mentioned in the Sect. 1 the road traffic network must be divided into required number of sub-networks prior the execution of its simulation in a distributed environment. The resulting sub-networks are then simulated as processes on particular nodes of the distributed computer (a sub-network per node). Since, even with the simulated road traffic network divided the vehicles still need to pass among the sub-networks. Thus, the simulation processes are interconnected by communication links enabling transfer of vehicles between the neighboring sub-networks in the form of messages (Potuzak 2012a). The inter-process communication is also necessary for synchronization of the simulation processes, which ensures that the simulation time of all processes is the same at the same moment (Potuzak 2012a).

There are two issues, which should be considered during the road traffic network division—the sub-networks load-balancing and the inter-process communication minimization. The load-balancing is necessary to achieve similar speeds of the simulation processes and, consequently, to minimize the waiting of the faster processes on the slower processes (Grosu et al. 2008; Potuzak 2012a). If the target distributed computer is a homogenous cluster (i.e. with nodes of the same computing power), the sub-networks should have similar loads (i.e. similar numbers of vehicles moving within them). If the target distributed computer is a heterogeneous cluster (i.e. with nodes of different computing power), the loads of the particular sub-networks should correspond to the computing powers of the nodes, on which the sub-networks will be simulated (Potuzak 2013a).

The communication minimization is necessary, because it is very slow in comparison to the remainder computations of the distributed simulation computations. The communication can be diminished by reducing the number of vehicles transferred among the sub-networks. This, in turn, can be achieved by minimization of the number of traffic lanes interconnecting the sub-networks (Potuzak 2013b).

3 Road Traffic Network Division Using DGA

The method for road traffic network division, which we have developed, considers both mentioned issues (see Sect. 2.2). There are two versions of the method—for homogenous and for heterogeneous clusters. The versions differ only in the load-balancing, where the road traffic network is divided into sub-networks with the similar loads for the homogeneous clusters and into sub-networks with the loads ratio corresponding to the computing power ratio of the particular nodes for the heterogeneous clusters (Potuzak 2013a).

Regardless the version, the method utilizes a (usually) less-detailed road traffic simulation for assigning of the weights to particular traffic lanes. These weights correspond to the numbers of vehicles moving within the lanes during the simulation run. It is possible to use a macroscopic, a mesoscopic, or a microscopic simulation. However, the macroscopic simulation is usually used, since all the simulations give similar results and the macroscopic simulation is the fastest one (Potuzak 2012a). Once the weights are assigned to the traffic lanes, the road traffic network is considered as a weighted graph. The crossroads are then acting as nodes and the sets of traffic lanes interconnecting neighboring crossroads acting as edges with weights equal to the sum of weights of all traffic lanes of each set (Potuzak 2013b). The road traffic network (i.e. the weighted graph) is then divided using the dividing genetic algorithm (DGA) into required number of load-balanced sub-networks of minimized number of edges (and, consequently, traffic lanes) interconnecting them (i.e. divided traffic edges/lanes) (Potuzak 2012a, 2013a).

3.1 Dividing Genetic Algorithm Description

The DGA is a standard genetic algorithm (see Sect. 2.1). Each individual represents a single assignment of crossroads to particular sub-networks in the form of a vector of integer values. The length of the vector corresponds to the total number of crossroads in the divided road traffic network. Each value then represents the sub-network to which the corresponding crossroad is assigned (Potuzak 2011).

The initial population of 90 individuals is randomly generated. So, the crossroads are randomly assigned to the particular sub-networks (Potuzak 2012a). The fitness value of each individual is calculated using the fitness function consisting of two parts—the *equability* and the *compactness*. The equability represents the load-balancing of the sub-networks and it is the only part of the DGA, which is different for homogenous and heterogeneous clusters. The compactness represents the minimization of the number of divided edges (and, consequently, traffic lanes) and is the same for both types of clusters. The fitness function can be then expressed as:

$$F_{DGA} = r_E \cdot E + (1 - r_E) \cdot C \tag{1}$$

where a F_{DGA} is the fitness value, E is the equability, C is the compactness, and r_E is the equability ratio determining, which part (the equability E or the compactness C) of the fitness function is more important. Different values of the r_E can be convenient in different situations. Usually, the r_E is set to 0.25.

Once the fitness values are calculated for all individuals, ten individuals are selected to be parents of the new generation using truncation selection (see Fig. 1a), which means than the individuals with the highest values from entire

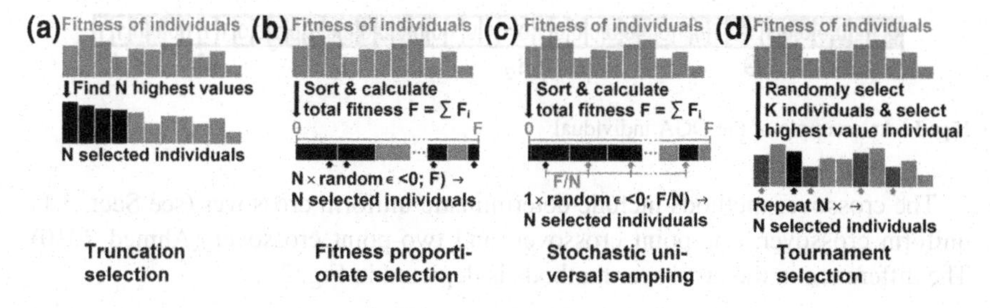

Fig. 1 Comparison of the implemented selection methods

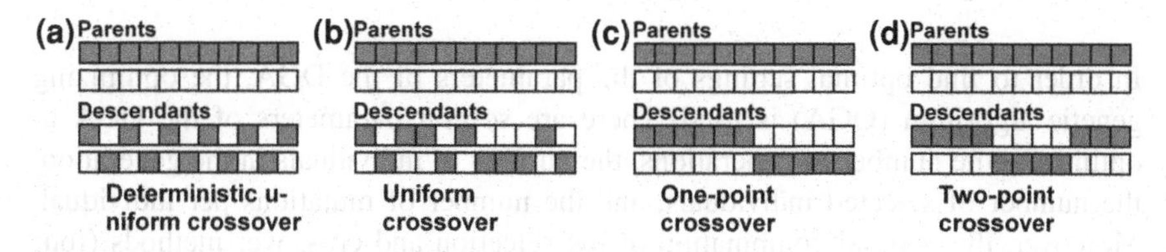

Fig. 2 Comparison of the implemented crossover methods

population are selected (Bäck 1996). Using all possible pairs formed from the selected individuals and the deterministic uniform crossover (see Fig. 2a), the new generation of 90 individuals is created. Using this crossover, every second value is exchanged between the two parents' vectors to form two descendants. Each descendant is then mutated using a preset number of random changes of its vector. This way, a new generation is created. The entire process repeats for the preset number of generations (up to 100,000) (Potuzak 2013b).

3.2 Implemented Improvements of DGA

In order to achieve a higher speed of the DGA, its implementation is recently refined. A new data structure is created to enable faster calculation of the fitness values and the utilization of integer instead of real numbers is employed where possible. This leads to a substantial increase of the DGA speed due to reducing its computation time to roughly 30 % of its original implementation.

Moreover, three other selection methods and three other crossover methods are implemented in order to include them to the OGA optimization of the DGA. A specific selection and crossover methods can be set prior to the DGA execution. The selection methods include truncation selection, fitness proportionate selection (Bäck 1996), stochastic universal sampling (Baker 1987), and tournament selection (Xie and Zhang 2013). The difference between the selection methods is depicted in Fig. 1.

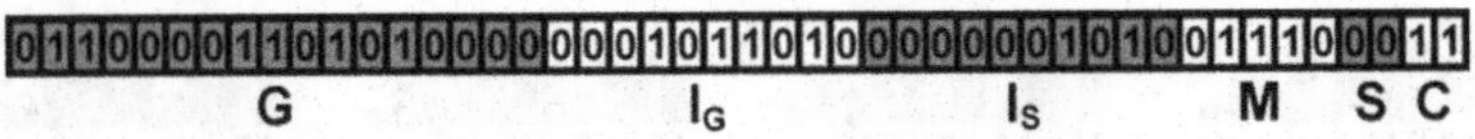

0110000110101000000010110100000001010011100011

G I_G I_S M S C

Fig. 3 An example of the OGA individual

The crossover methods include deterministic uniform crossover (see Sect. 3.1), uniform crossover, one-point crossover, and two-point crossover (Ahmed 2010). The difference in the crossover methods is depicted in Fig. 2.

4 DGA Optimization Using OGA

In order to find optimal settings of the parameters of the DGA, the optimizing genetic algorithm (OGA) is used. There are several parameters of the DGA to optimize—the number of generations, the number of individuals in the generation, the number of selected individuals, and the number of mutations per individual. Moreover, the optimal combination of the selection and crossover methods (four and four options, respectively—see Sect. 3.2) should be determined. All the mentioned parameters can significantly influence the quality of road traffic network division and the speed of the division method (Potuzak 2013b).

A parameter, which will not be optimized using the OGA is the equability ratio (r_E), which determines whether the load-balancing of the sub-networks or the minimization of the number of divided traffic lanes should be more important during the division (see Sect. 3.1). Since the requirements can vary in different situations, it would be difficult to find a single optimal value (Potuzak 2013b).

4.1 Optimizing Genetic Algorithm Description

The utilization of the OGA for optimization of the DGA seems to be a viable approach, since the genetic algorithms are generally convenient for optimization problems (Farshbaf and Feizi-Darakhshi 2009). The OGA individual representation must incorporate all the parameters, which shall be optimized (see Sect. 4). These parameters can be expressed as nonnegative integer numbers of different sizes. So, the OGA individual is represented by a binary vector with several multi-bit parts. Each part represents a single DGA parameter to optimize encoded using the unsigned magnitude representation. More specifically, each individual consists of 46 bits—17 bits for the number of generations (G), 10 bits for the number of individuals in a generation (I_G), 10 bits for the number of selected individuals (I_S), 5 bits for the number of mutations per individual (M), 2 bits for the type of selection (S), and 2 bits for the type of crossover (C) (Potuzak 2013b). The example is depicted in Fig. 3.

The initial population of the OGA consists of 90 randomly generated individuals (similarly to DGA). For all individuals, the fitness value is calculated. For an OGA individual, it is necessary to perform the complete DGA on a road traffic network with parameters set according to the OGA individual. Since it is necessary for the OGA optimization to be universal and not solely for one road traffic network, three road traffic networks of different sizes divided into three different numbers of sub-networks will be used. These networks are regular square grids of 64, 256, and 1,024 crossroads (86, 326, and 1,267 km of total traffic lanes length, respectively). They are divided into 2, 4, and 8 sub-networks. Originally, the division into 16 sub-networks was considered as well (Potuzak 2013b). However, for the road traffic networks of the given sizes, the division into 16 sub-networks is quite improbable to be convenient, because the resulting sub-networks will be too small.

The OGA fitness value is then calculated using the maximal achieved DGA fitness values of all combinations of the divided road traffic networks and the number of sub-networks. So, the fitness function can be expresses as:

$$F_{OGA} = \frac{\sum_{i=1}^{3}\sum_{j=1}^{3} F_{ij}^{\max}}{9}, \tag{2}$$

where F_{OGA} is the OGA fitness function and $F_{ij}^{\max}$is the maximal achieved DGA fitness value for ith road traffic network and jth number of sub-networks.

Once the fitness values are calculated for all OGA individuals, ten individuals are selected to be parents of the new generation. The truncation selection is used (see Fig. 1a), which means that the individuals with the highest fitness values are selected. For the creation of the descendants, the deterministic uniform crossover (see Fig. 2a) of all possible pairs of two parents is used. Each descendant can be mutated using random but limited number of random changes of its bits. Once the new generation is created, the fitness values are calculated for all OGA individuals and the process repeats until the preset number of generations is created. So, the OGA is very similar to the original DGA.

4.2 Distributed OGA Description

Because the OGA is expected to be extremely computation intensive (Potuzak 2012b, 2013b), only its distributed implementation seems to be feasible. Still a distributed computer with large number of nodes will be required to keep the OGA computation time in acceptable limits. For the OGA, we can utilize two classrooms, which are at our disposal at Department of Computer Sciences and Engineering of University of West Bohemia (DSCE UWB). Each classroom contains twelve desktop PCs. Each PC incorporates Intel i5-2400S Quad-Core processor at 3.1 GHz, 8 GB RAM, and 250 GB HDD. The PCs are used for the education purposes during the working days, but are usually idle or deactivated during the late afternoons, nights, and weekends. As these times, they can be readily used for the OGA computations (Potuzak 2013b).

For the distributed OGA, the well-known farmer–worker paradigm is used. So, there is one control process (farmer) and a number of working processes

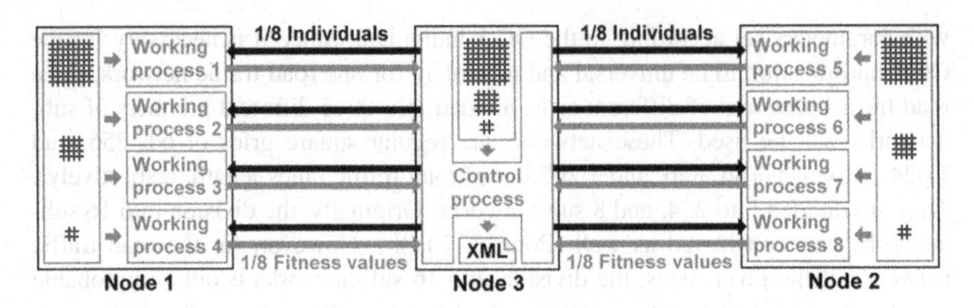

Fig. 4 The OGA communication scheme for one control and eight working processes

(workers). The OGA is implemented in the DUTS Editor system (developed at DSCE UWB). This system with all three road traffic networks (see Sect. 4.1) must be installed on each computer, which will participate on the OGA computation. It is possible to run either the control process or the working process. It is convenient, on a multi-core node (computer), to run number of working processes corresponding to the number of processor cores. Each working process is connected by a bidirectional communication link to the control process for the message passing.

Assume now that the distributed OGA is performed on three quad-core nodes, one of them hosting the control process while two other hosting four working processes each (see Fig. 4). The OGA computation is then performed as follows. Each working process loads all three road traffic networks. The control process loads them as well and performs the macroscopic simulation on each of them in order to obtain weights of the traffic lanes (i.e. input for the DGA—see Sect. 3). These weights are sent to all working processes, which assign them to the corresponding traffic lanes of the loaded road traffic networks. The control process then creates the initial OGA population and distributes the individuals uniformly among the working process. Each working process then calculates the OGA fitness values of the received individuals by performing nine DGA runs per OGA individual (i.e. using three road traffic networks divided into three different numbers of sub-networks—see Sect. 4.1). The OGA fitness values are then sent to the control process. Once the control process receives the fitness values of all OGA individuals, it performs the selection, crossover, and mutation and creates a new OGA generation.

Its OGA individuals are again distributed to the working processes and the entire process repeats for the preset number of generations. The inter-process communication scheme of the OGA is depicted in Fig. 4.

Because the fitness values calculation is by far the most computation-intensive part of the OGA and only relatively short vectors of integer or real values are transferred two times per generation between the control and each working process, the time necessary for inter-process communication is negligible, which is a very convenient feature for a distributed computation. Nevertheless, the OGA computation time is in range of days and weeks (see Sect. 5) and the classrooms intended for the OGA are regularly used for education purposes. Hence, a single continuous computation of the OGA is out of question. Instead, it is possible for the control process to store the weights of the traffic lanes and the current generation in a XML file during the computation. So, the computation can be ended whenever necessary and then resumed using the data in the XML file.

5 Tests and Results

Currently, an implementation of the distributed OGA is nearly finished. In order to determine the time necessary for its execution a set of tests have been performed using one processor core of one node (computer) of the distributed computer, on which the distributed OGA is performed (see Sect. 4.2).

The set of tests has been focused on the time necessary for the homogenous and heterogeneous DGA runs in dependence on the settings of its three optimized parameters. The settings of these parameters—the number of DGA generations, the DGA selection type, and the DGA crossover type (see Sect. 4) can significantly influence the time necessary for the DGA run, and consequently, the distributed OGA execution, which utilize nine DGA runs for the computation of the fitness value of a single OGA individual (see Sect. 4.1). The tests have been performed for all three road traffic networks (see Sect. 4.1), 1,000, 10,000, and 100,000 DGA generations, and all combinations of the DGA selection and DGA crossover types. All road traffic networks have been divided only into four sub-networks, since the number of sub-networks has negligible effect on the DGA computation time (Potuzak 2013b). The results for the DGA for homogeneous cluster are summarized in Table 1. Each value is averaged from ten attempts. The results for the DGA for heterogeneous clusters are very similar and differ in range of several percents. Therefore, from now on, we will not distinguish between these two versions of the DGA.

The results depicted in Table 1 can be used for the determination of the mean computation time of the fitness value of a single OGA individual using one processor core depending on the number of DGA generations. Following the calculation of the OGA fitness function (see Eq. 2) and using the last row of Table 1, this mean time can be calculated as:

$$\overline{T_G^{OGA}} = 3 \cdot \left(\overline{T_{64,G}^{DGA}} + \overline{T_{256,G}^{DGA}} + \overline{T_{1,024,G}^{DGA}} \right), \tag{3}$$

where $\overline{T_G^{OGA}}$ is the mean computation time necessary for the calculation of the OGA individual for G DGA generations, and $\overline{T_{X,G}^{DGA}}$ are the mean times necessary for the DGA for G generations and X crossroads of the divided road traffic network (i.e. values in last row of Table 1). It should be noted that the distributed OGA is intended to utilize all four processor cores of each node of the distributed computer. It has been determined in (Potuzak 2012b) that it is possible for the price of 7 % slowdown in comparison to the utilization of only one core. The mean computation time of the fitness value of a single OGA individual for different numbers of DGA generations is summarized in Table 2.

Table 1 Time necessary for the DGA for homogeneous clusters

Crossroads		64			256			1,024		
Generations		10^3	10^4	10^5	10^3	10^4	10^5	10^3	10^4	10^5
S	C	DGA computation time (s)								
TR	DU	0.2	2.2	22.3	0.7	6.2	58.4	2.8	24.2	211.8
	U	0.3	2.8	28.8	0.9	8.5	81.4	3.7	33.4	325.4
	1P	0.2	2.2	22.4	0.7	6.1	58.3	2.7	23.7	207.2
	2P	0.2	2.2	22.7	0.7	6.2	59.6	2.8	24.1	215.4
FP	DU	0.2	2.6	26.9	0.8	7.6	75.8	2.8	27.8	764.0
	U	0.3	3.2	33.3	1.0	10.1	100.9	3.8	37.8	394.3
	1P	0.2	2.6	26.4	0.7	7.5	74.5	2.8	27.2	271.0
	2P	0.3	2.6	26.8	0.8	7.5	75.3	2.8	27.6	273.9
SUS	DU	0.2	2.6	26.4	0.8	7.5	75.1	2.8	27.7	275.2
	U	0.3	3.2	32.8	1.0	10	100.3	3.8	37.8	393.7
	1P	0.2	2.5	26.1	0.7	7.4	73.9	2.8	27.2	270.2
	2P	0.2	2.5	26.3	0.7	7.5	74.7	2.8	27.4	273.1
TO	DU	0.2	2.2	22.6	0.7	6.2	58.9	2.8	24.4	213.6
	U	0.3	2.8	28.8	0.9	8.6	82.2	3.7	34.0	326.2
	1P	0.2	2.1	22.3	0.7	6.0	57.8	2.8	23.8	209.1
	2P	0.2	2.2	22.6	0.7	6.1	58.5	2.8	23.9	211.8
Average		0.2	2.5	26.1	0.8	7.4	72.8	3.0	28.2	302.2

Selection (S) types: *TR* truncation selection, *FP* fitness proportionate selection, *SUS* stochastic universal sampling, *TO* tournament selection. Crossover types: *DU* deterministic uniform crossover, *U* uniform crossover, *1P* one-point crossover, *2P* two-point crossover

Table 2 The mean computation time of the fitness value of a single DGA individual

	Without 7 % slowdown	With 7 % slowdown
DGA generations count	Computation time (s)	
10^3	12.0	12.8
10^4	114.3	122.3
10^5	1,203.3	1,287.5
Average	443.2	474.2

The mean computation time of the fitness value of a single OGA individual for mean number of generations increased by 7 % (see most right cell in last row of the Table 2) can be designated the *OGA individual computation time*. Comparing the new results depicted in Table 2 with the results described in Potuzak (2013b), the new OGA individual computation time is only roughly 17 % of the OGA individual computation time described in Potuzak (2013b). The reasons of this speedup are the optimization of the DGA (see Sect. 3.2) and the change of the calculation of the OGA fitness function (the division into 16 sub-networks is no longer considered—see Sect. 4.1).

Using the OGA individual computation time and knowing there are 90 individuals per OGA generation, it is possible to determine the total computation time of the OGA. Considering that the distributed computer for the OGA would consist of

Table 3 Total computation time of the OGA depending on the number of OGA generations

OGA generations	Computation time (h)	Computation time (days)
10^3	123.496	5.146
10^4	1,234.958	51.457
10^5	12,349.583	514.566

24 nodes (i.e. computers from both considered classrooms—see Sect. 4.2), there are $24 \times 4 = 96$ processors' cores available for the OGA working processes. The OGA control process can reside in another computer at our disposal. The OGA computation time in dependence on the number of OGA generations is summarized in the Table 3.

As can be seen in the Table 3, the OGA computation time ranges from 5 to 515 days, depending on the number of OGA generations. So, at least the lower numbers of OGA generations seem to be feasible. The time necessary for the macroscopic simulation used for the obtaining of the weights of traffic lanes is not included in the OGA computation time. The reason is that the macroscopic simulation is performed only once and is very fast (several seconds for the largest road traffic network). Similarly, the time necessary for inter-process communication is not included, because the communication is performed rarely during the OGA computations (two messages per OGA generation per working process).

6 Conclusions

In this paper, we described the optimization of the genetic algorithm for road traffic network division (DGA) using a distributed optimizing genetic algorithm (distributed OGA). We focus on the determination of the total time necessary for the distributed OGA computation. Depending on the number of OGA generations, this computation time ranges from 5 to 515 days on 24 quad-core computers. So, the OGA seems to be feasible at least for lower numbers of OGA generations.

Currently, the OGA implementation is nearly complete. The debugging and preliminary testing, which are crucial for such a long computation, are now ongoing. The further step in our research is to perform the OGA in order to find optimal settings of the DGA parameters.

References

Ahmed ZH (2010) Genetic algorithm for the traveling salesman problem using sequential constructive crossover operator. Int J Biom Bioinform 3(6):96–105

Bäck T (1996) Evolutionary algorithms in theory and practice: evolution strategies, evolutionary programming, genetic algorithms. Oxford University Press, New York

Baker JE (1987) Reducing bias and inefficiency in the selection algorithm. In: Proceedings of the second international conference on genetic algorithms and their application, pp 14–21

Farshbaf M, Feizi-Darakhshi M (2009) Multi-objective optimization of graph partitioning using genetic algorithms. In: 2009 third international conference on advanced engineering computing and applications in sciences, Sliema, pp 1–6

Grosu D, Chronopoulos AT, Leung MY (2008) Cooperative load balancing in distributed systems. Concurr Comput Pract Exp 20(16):1953–1976

Menouar B (2010) Genetic algorithm encoding representations for graph partitioning problems. In: 2010 international conference on machine and web intelligence, Algiers, pp 288–291

Poli R, Langdon WB, McPhee NF (2008) A field guide to genetic programming. Published via http://lulu.com and freely available at http://www.gp-field-guide.org.uk (with contributions by Koza JR)

Potuzak T (2011) Suitability of a genetic algorithm for road traffic network division. In: KDIR 2011—proceedings of the international conference on knowledge discovery and information retrieval, Paris, pp 448–451

Potuzak T (2012a) Methods for division of road traffic networks focused on load-balancing. Adv Comput 2(4):42–53

Potuzak T (2012b) Issues of optimization of a genetic algorithm for traffic network division using a genetic algorithm. In: Proceedings of the international conference on knowledge discovery and information retrieval, Barcelona, pp 340–343

Potuzak T (2013a) Methods for division of road traffic network for distributed simulation performed on heterogeneous clusters. Comput Sci Inf Syst 10(1):321–348

Potuzak T (2013b) Feasibility study of optimization of a genetic algorithm for traffic network division for distributed road traffic simulation. In: Proceedings of the 6th international conference on human system interaction, pp 372–379

Xie H, Zhang M (2013) Parent selection pressure auto-tuning for tournament selection in genetic programming. IEEE Trans Evol Comput 17(1):1–18

Optimal Design of Rule-Based Systems by Solving Fuzzy Relational Equations

A. P. Rotshtein and H. B. Rakytyanska

Abstract In this chapter, an approach to the design of rule-based systems within the framework of fuzzy relational calculus is proposed. The system of fuzzy relations serves as the generator of the rule-based solutions of fuzzy relational equations. Each solution represents a different trade-off between the classification accuracy and the number of fuzzy rules. The accuracy-complexity trade-off is achieved by optimization of the total number of decision classes for relations and rules.

1 Introduction

In the design of fuzzy rule-based systems we have two conflicting objectives: accuracy maximization and complexity minimization. For the last years, the problem of finding the right accuracy-complexity trade-off has given rise to a growing interest in neural and genetic systems that take both aspects into account.

Radial basis function (RBF) neural networks serve as the universal platform for rule extraction (Fu and Wang 2001). Rule extraction from RBF networks combined with trained SVMs consists of constructing hyperboxes using support vectors (Zhang et al. 2008). The general fuzzy min-max (GFMM) neural networks form the decision boundaries by covering the pattern space with hyperboxes (Gabrys and Bargiela 2000). The solution of the problem of neural network structure optimization involves finding a compromise between the generalization ability and the ability to capture nonlinear boundaries between classes. The specific training mode in the min-max neural networks consists of reducing the number of hyperboxes

A. P. Rotshtein (✉)
Jerusalem College of Technology—Machon Lev, Jerusalem, Israel
e-mail: rot@jct.ac.il

H. B. Rakytyanska
Vinnitsa National Technical University, Vinnitsa, Ukraine
e-mail: h_rakit@ukr.net

Z. S. Hippe et al. (eds.), *Issues and Challenges in Artificial Intelligence*,
Studies in Computational Intelligence 559, DOI: 10.1007/978-3-319-06883-1_14,

admitting no loss of recognition performance. However, both goals can be achieved by specifying the maximum hyperbox size, which defines how many hyperboxes should be created. The undesirable effects of hyperbox expansion are overlapping hyperboxes. In this case, one pattern fully belongs to two or more decision classes.

The straightforward approach to the design of genetic rule-based systems supposes rule selection. The multi-objective rule selection has been applied to improve the accuracy-complexity trade-off. Genetic refinement systems select the alternative refinements by using rule evaluation measures to cover some training set and avoid misclassification errors. Similarity measures are applied to detect and merge compatible fuzzy sets and to remove "don't-care" terms (Jin 2000). Confidence and support measures are applied to generate the candidate rules and select the alternative refinement (Ishibuchi and Yamamoto 2004). Finally, these criteria are incorporated into multi-objective genetic algorithms in order to simultaneously achieve the accuracy maximization and the complexity minimization.

In this chapter, we propose an approach to the design of rule-based systems within the framework of fuzzy relational calculus (Peeva and Kyosev 2004). In paper (Rotshtein and Rakytyanska 2013) we used the system of fuzzy relations as the generator of the rule-based solutions of fuzzy relational equations. The system of fuzzy IF–THEN rules can be rearranged as a collection of linguistic solutions of fuzzy relational equations using the composite system of fuzzy terms. It allows us to avoid alternative rule selection and eliminate overlaps between classes. The solution of fuzzy relational equations guarantees the optimal number of fuzzy rules for each output fuzzy term and the optimal geometry of input fuzzy terms for each linguistic solution. Each solution of fuzzy relational equations represents a different trade-off between the classification accuracy and the number of fuzzy rules. In this case, the accuracy-complexity trade-off can be achieved by including the optimization of the total number of decision classes for relations and rules.

The approach proposed is illustrated by the example of the design of the inventory control system.

2 Fuzzy Model of the Object

Let us consider an object $y = f(\mathbf{X})$ with n inputs $\mathbf{X} = (x_1, \ldots, x_n)$ and one output y, for which the inputs–output interconnection can be represented in the form of the following system of IF–THEN rules:

$$\bigcup_{p=\overline{1,z_j}} \left[\bigcap_{i=\overline{1,n}} \left(x_i = a_i^{jp} \right) \text{ with weight } w_{jp} \right] \rightarrow y = d_j, \quad j = \overline{1,m}, \tag{1}$$

where a_i^{jp} is the term for variable x_i evaluation in the row with number $jp, j = \overline{1,m}, p = \overline{1,z_j}$; d_j is the term for variable y evaluation; z_j is the number of rules corresponding to the term d_j; m is the number of output terms; w_{jp} is the weight of the rule with number jp.

This knowledge base can be rearranged as a collection of linguistic solutions of fuzzy relational equations using the composite system of fuzzy terms.

Let $\{C_1,\ldots,C_N\} = \{c_{i1},\ldots,c_{ik_i}\}, i=\overline{1,n}$, and $\{D_1,\ldots,D_M\}$ denote the sets of first level terms for variables x_i and y evaluation; $\left\{u_1^I,\ldots,u_{K_I}^I\right\}$, $I=\overline{1,N}$, $N=k_1+\cdots+k_n$, and $\left\{v_1^J,\ldots,v_{Q_J}^J\right\}$, $J=\overline{1,M}$, denote the sets of second level terms for significance measures μ^{C_I} and μ^{D_J} evaluation. Second level terms are associated with intervals $u_P^I \in \left[\underline{\mu}_P^{C_I},\overline{\mu}_P^{C_I}\right], P=\overline{1,K_I}$, and $v_L^J \in \left[\underline{\mu}_L^{D_J},\overline{\mu}_L^{D_J}\right]$, $L=\overline{1,Q_J}$, where $\underline{\mu}_P^{C_I}\left(\overline{\mu}_P^{C_I}\right)$ and $\underline{\mu}_L^{D_J}\left(\overline{\mu}_L^{D_J}\right)$ are the lower (upper) bounds of significance measures $\mu^{C_I}=u_P^I$ and $\mu^{D_J}=v_L^J$. Pairs $(c_{il},\mu^{c_{il}}=u_P^{il})$ and $(D_J,$ $\mu^{D_J}=v_L^J)$ are associated with the composite fuzzy terms U_P^{il} and V_L^J for variables x_i and y evaluation. Let $\{a_1,\ldots,a_q\}=\left\{U_1^I,\ldots,U_{K_I}^I\right\}$ and $\{d_1,\ldots,d_m\}=$ $\left\{V_1^J,\ldots,V_{Q_J}^J\right\}$ redenote the sets of terms for variables x_i and y evaluation, where $q=K_1+\cdots+K_N$ and $m=Q_1+\cdots+Q_M$ are the total numbers of input and output terms in (1).

Thus, the knowledge base (1) is equivalent to the following system of fuzzy rules, which interconnects the composite input and output terms:

$$\bigcup_{p=\overline{1,z_j}}\left[\bigcap_{i=\overline{1,n}}\left(x_i=A_i^{jp}\bigcap\mu^{A_i^{jp}}=\alpha_i^{jp}\right)\text{ with weight } w_{jp}\right]\to y=d_j,\quad j=\overline{1,m}, \tag{2}$$

where A_i^{jp} is the first level term for variable x_i evaluation; α_i^{jp} is the second level term for the significance measure $\mu^{A_i^{jp}}$ evaluation.

The composite knowledge base (2) can be decomposed into the separate knowledge bases for relations and rules. The collection of fuzzy relational matrices $\mathbf{R}_i\subseteq c_{il}\times D_J=\left[r_{il}^J\right], i=\overline{1,n}, l=\overline{1,k_i},\quad J=\overline{1,M}$, interconnects the first level input and output terms using the following system of fuzzy relational rules:

$$\bigcap_{i=\overline{1,n}}\left\{\bigcup_{l=\overline{1,k_i}}\left(x_i=c_{il}\text{ with weight } r_{il}^J\right)\right\}\to y=D_J,\quad J=\overline{1,M}. \tag{3}$$

The following system of fuzzy rules interconnects the second level input and output terms:

$$\bigcup_{p=\overline{1,z_j}}\left[\bigcap_{i=\overline{1,n}}\left\{\bigcup_{l=\overline{1,k_i}}\left(\mu^{c_{il}}=v_{il}^{jp}\right)\right\}\text{ with weight } w_{jp}\right]\to y=d_j,\quad j=\overline{1,m}, \tag{4}$$

where $v_{il}^{jp}=\alpha_i^{jp}$ if $A_i^{jp}=c_{il}$, otherwise, $v_{il}^{jp}=0$.

The inputs–output dependency can be described with the help of the system of fuzzy logic equations for relations and rules (Yager and Filev 1994):

$$\mu^{D_J}(y)=\min_{i=\overline{1,n}}\left[\max_{l=\overline{1,k_i}}\left(\min\left(\mu^{c_{il}}(x_i),r_{il}^J\right)\right)\right],\quad J=\overline{1,M}, \tag{5}$$

$$\mu^{d_j}(y) = \max_{p=\overline{1,z_j}} \left\{ w_{jp} \cdot \min_{i=\overline{1,n}} \left[\mu^{a_i^{jp}}(x_i) \right] \right\}, \quad j = \overline{1,m}, \tag{6}$$

where $\mu^{D_J}(y)$ and $\mu^{c_{il}}(x_i)$ are membership functions of variables y and x_i to the fuzzy terms D_J and c_{il}; $\mu^{d_j}(y)$ and $\mu^{a_i^{jp}}(x_i)$ are membership functions of variables y and x_i to the fuzzy terms d_j and a_i^{jp}.

We use a bell-shaped membership function model of variable u to term T:

$$\mu^{T}(u) = 1 \Big/ \left(1 + ((u - \beta)/\sigma)^2 \right), \tag{7}$$

where β is a coordinate of maximum; σ is a parameter of concentration.

For relations and rules, the operation of defuzzification is defined as follows:

$$y = \sum_{J=1}^{M} \underline{y}_J^D \cdot \mu^{D_J}(y) \Big/ \sum_{J=1}^{M} \mu^{D_J}(y), \tag{8}$$

$$y = \sum_{j=1}^{m} \underline{y}_j^d \cdot \mu^{d_j}(y) \Big/ \sum_{j=1}^{m} \mu^{d_j}(y), \tag{9}$$

where $\underline{y}_J^D$ and $\underline{y}_j^d$ are the bounds of decision classes D_J and d_j.

Correlations (5)–(9) define the generalized fuzzy model of the object approximated by relations and rules as follows:

$$y = f_R(\mathbf{X}, M, \mathbf{R}, \mathbf{B}_C, \mathbf{\Omega}_C),$$
$$y = f_r(\mathbf{X}, f_R, m, Z, q, \mathbf{W}, \mathbf{B}_a, \mathbf{\Omega}_a),$$

where $\mathbf{B}_C = \left(\beta^{C_1}, \ldots, \beta^{C_N}\right)$ and $\mathbf{\Omega}_C = \left(\sigma^{C_1}, \ldots, \sigma^{C_N}\right)$ are the vectors of parameters for fuzzy terms C_I membership functions; $\mathbf{W} = (w_1, \ldots, w_Z)$ is the vector of rules weights; Z is the total number of rules; $\mathbf{B}_a = (\beta^{a_1}, \ldots, \beta^{a_q})$ and $\mathbf{\Omega}_a = (\sigma^{a_1}, \ldots, \sigma^{a_q})$ are the vectors of parameters for fuzzy terms a_k membership functions; F_R and F_r are the operators of inputs–output connection, corresponding to formulae (5), (7), (8), and formulae (6), (7), (9), respectively.

3 Optimization Problem Statement

Let us define the training data as a set of L pairs: $\langle \widehat{\mathbf{X}}_s, \hat{y}_s \rangle, s = \overline{1,L}$, where $\widehat{\mathbf{X}}_s = \left(\hat{x}_1^s, \ldots, \hat{x}_n^s\right)$ and $\hat{y}_s$ are the input vector and its corresponding output in the experiment number s. We shall evaluate the complexity of the fuzzy system using the total number $Z(M, m)$ of rule-based solutions (4) of fuzzy relational equations (5). We shall evaluate the performance of the fuzzy system using the following root mean-squared error:

$$E = \sqrt{\frac{1}{L} \sum_{s=1}^{L} [f_r(\widehat{\mathbf{X}}_s, M, m) - \hat{y}_s]^2}$$

The optimization problem can be formulated as follows.

Direct statement It is required to find M and m, for which $Z(M,\ m) \rightarrow \min$ and $E(M,\ m) \leq \overline{E}$, where $\overline{E}$ is the permissible root mean-squared error.

Dual statement It is required to find M and m, for which $E(M,m) \rightarrow \min$ and $Z(M,m) \leq \overline{Z}$, where $\overline{Z}$ is the permissible number of rules.

The gradient algorithm (Rotshtein and Kuznetcov 1992) is proposed for finding the total number of classes for relations and rules. Given the number of classes M and m, the genetic and neural algorithm is used for structural and parametric tuning of relations and rules (Rotshtein and Rakytyanska 2012).

4 Algorithm of Optimization

We shall denote: $\mathbf{\Psi}(M, m, \mathbf{\Psi}_R, \mathbf{\Psi}_r)$ is the vector of the fuzzy model parameters, where $\mathbf{\Psi}_R = (\mathbf{R}, \mathbf{B}_C, \mathbf{\Omega}_C)$ and $\mathbf{\Psi}_r(\mathbf{W}, \mathbf{B}_a, \mathbf{\Omega}_a)$ are the vectors of parameters for relations and rules, respectively.

Gradient γ_ψ is defined as the following ratio: $\gamma_\psi = \Delta E(\mathbf{\Psi}, \psi) / \Delta Z(\mathbf{\Psi}, \psi)$, where $\Delta E(\mathbf{\Psi}, \psi) = E(\mathbf{\Psi}, \psi) - E(\mathbf{\Psi}, \psi + 1)$ and $\Delta Z(\mathbf{\Psi}, \psi) = Z(\mathbf{\Psi}, \psi + 1) - Z(\mathbf{\Psi}, \psi)$ are the error decrease and the rules number increase while the number of output classes ψ is increasing. The coordinate ψ is interpreted as M or m.

Algorithm for the direct problem solving will have the following form.
Let $\mathbf{\Psi}^{(k)} = \left(M^{(k)}, m^{(k)}, \mathbf{\Psi}_R^{(k)}, \mathbf{\Psi}_r^{(k)}\right)$ be some kth solution.

Step 1 Initialize the null variant of the fuzzy model: $k := 0, \mathbf{\Psi}^{(0)} = \left(M^{(0)}, m^{(0)}, \mathbf{\Psi}_R^{(0)}, \mathbf{\Psi}_r^{(0)}\right)$. If $E\left(\mathbf{\Psi}^{(0)}\right) < \overline{E}$, then go to Step 5.

Step 2 If $E(\mathbf{\Psi}^{(k)}) > \overline{E}$, then go to Step 3, otherwise, go to Step 4.

Step 3 For the models $\mathbf{\Psi}' = \left(M^{(k)} + 1, m^{(k)}, \mathbf{\Psi}'_R, \mathbf{\Psi}'_r\right)$ and $\mathbf{\Psi}'' = \left(M^{(k)}, m^{(k)} + 1, \mathbf{\Psi}''_R, \mathbf{\Psi}''_r\right)$, determine the gradients γ_M and γ_m. Find the coordinate, for which $\gamma = \max\{\gamma_M, \gamma_m\}, k := k + 1$. If $\gamma = \gamma_M$, then $\mathbf{\Psi}^{(k)} := \mathbf{\Psi}'$. If $\gamma = \gamma_m$, then $\mathbf{\Psi}^{(k)} := \mathbf{\Psi}''$. Go to Step 2.

Step 4 Refine the model $\mathbf{\Psi}^{(k)}$. For the models $\mathbf{\Psi}' = \left(M^{(k-1)}, m^{(k-1)} + 1, \mathbf{\Psi}'_R, \mathbf{\Psi}'_r\right)$ and $\mathbf{\Psi}'' = \left(M^{(k-1)} + 1, m^{(k-1)}, \mathbf{\Psi}''_R, \mathbf{\Psi}''_r\right)$, check the conditions

$$E(M^{(k-1)}, m^{(k-1)} + 1) \leq \overline{E}, \quad \text{if } \gamma = \gamma_M; \tag{10}$$

$$E(M^{(k-1)} + 1, m^{(k-1)}) \leq \overline{E}, \quad \text{if } \gamma = \gamma_m. \tag{11}$$

If conditions (10) and (11) are not satisfied, then $\mathbf{\Psi}^{(k)}$ is the resultant solution, otherwise, find $Z\left(\mathbf{\Psi}'\right)$ and $Z(\mathbf{\Psi}'')$. If $Z(\mathbf{\Psi}') < Z(\mathbf{\Psi}^{(k)})$, then $\mathbf{\Psi}'$ is the resultant solution. If $Z(\mathbf{\Psi}'') < Z(\mathbf{\Psi}^{(k)})$, then $\mathbf{\Psi}''$ is the resultant solution.

Step 5 Reduce the model $\mathbf{\Psi}^{(k)}$. For the models $\mathbf{\Psi}' = (M^{(k)} - 1, m^{(k)}, \mathbf{\Psi}'_R, \mathbf{\Psi}'_r)$ and $\mathbf{\Psi}'' = (M^{(k)}, m^{(k)} - 1, \mathbf{\Psi}''_R, \mathbf{\Psi}''_r)$, check the conditions:

$$E(M^{(k)} - 1, m^{(k)}) \leq \overline{E}; \tag{12}$$

$$E(M^{(k)}, m^{(k)} - 1) \leq \overline{E}. \tag{13}$$

If conditions (12) and (13) are not satisfied, then $\Psi^{(k)}$ is the resultant solution, otherwise, go to Step 6.

Step 6 For the coordinates, which satisfy conditions (12) and (13), find the decrease of the rules number $\Delta Z(\Psi, \psi)$. Find the coordinate, for which

$$\Delta = \max\{\Delta Z(\Psi, M^{(k)} - 1, m^{(k)}), \Delta Z(\Psi, M^{(k)}, m^{(k)} - 1)\},$$

$k := k + 1$. If $\Delta = \Delta Z(\Psi, M)$, then $\Psi^{(k)} := \Psi'$. If $\Delta = \Delta Z(\Psi, m)$, then $\Psi^{(k)} := \Psi''$. Go to Step 5.

Algorithm for the dual problem solving will have the following form.

Step 1 Initialize the null variant of the fuzzy model: $k := 0, \Psi^{(0)} = \left(M^{(0)}, m^{(0)}, \Psi_R^{(0)}, \Psi_r^{(0)}\right)$. If $Z\left(\Psi^{(0)}\right) > \overline{Z}$, then go to Step 6.

Step 2 If $Z\left(\Psi^{(k)}\right) < \overline{Z}$, then go to Step 3, otherwise, go to Step4.

Step 3 Coincides with Step 3 of the previous algorithm. Go to Step 2.

Step 4 Refine the model $\Psi^{(k)}$. For the models $\Psi' = \left(M^{(k-1)}, m^{(k-1)} + 1, \Psi_R', \Psi_r'\right)$ and $\Psi'' = \left(M^{(k-1)} + 1, m^{(k-1)}, \Psi_R'', \Psi_r''\right)$ check the conditions:

$$Z(M^{(k-1)}, m^{(k-1)} + 1) \leq \overline{Z}, \quad \text{if } \gamma = \gamma_M; \tag{14}$$

$$Z(M^{(k-1)} + 1, m^{(k-1)}) \leq \overline{Z}, \quad \text{if } \gamma = \gamma_m. \tag{15}$$

Step 5 If condition (14) or (15) is not satisfied, then $\Psi^{(k-1)}$ is the resultant solution, otherwise, $k := k + 1$. If $\gamma = \gamma_M$, then $\Psi^{(k)} := \Psi'$. If $\gamma = \gamma_m$, then $\Psi^{(k)} := \Psi''$. Go to Step 4.

Step 6 Reduce the model $\Psi^{(k)}$. For the models $\Psi' = \left(M^{(k)} - 1, m^{(k)}, \Psi_R', \Psi_r'\right)$ and $\Psi' = \left(M^{(k)}, m^{(k)} - 1, \Psi_R'', \Psi_r''\right)$, check the conditions

$$Z(M^{(k)} - 1, m^{(k)}) \leq \overline{Z}; \tag{16}$$

$$Z(M^{(k)}, m^{(k)} - 1) \leq \overline{Z}. \tag{17}$$

If conditions (16) and (17) are satisfied, then find $E\left(\Psi'\right)$ and $E\left(\Psi''\right)$, otherwise, go to Step 7. If $E\left(\Psi'\right) < E\left(\Psi''\right)$, then Ψ' is the resultant solution, otherwise, Ψ'' is the resultant solution.

Step 7 For the coordinates, which do not satisfy conditions (16) and (17), find the increase of the error $\Delta E(\Psi, \psi)$. Find the coordinate, for which

$$\Delta = \min\left\{\Delta E\left(\Psi, M^{(k)} - 1, m^{(k)}\right), \Delta E\left(\Psi, M^{(k)}, m^{(k)} - 1\right)\right\},$$

$k := k + 1$. If $\Delta = \Delta E(\Psi, M)$, then $\Psi^{(k)} := \Psi'$. If $\Delta = \Delta E(\Psi, m)$, then $\Psi^{(k)} := \Psi''$. Go to Step 6.

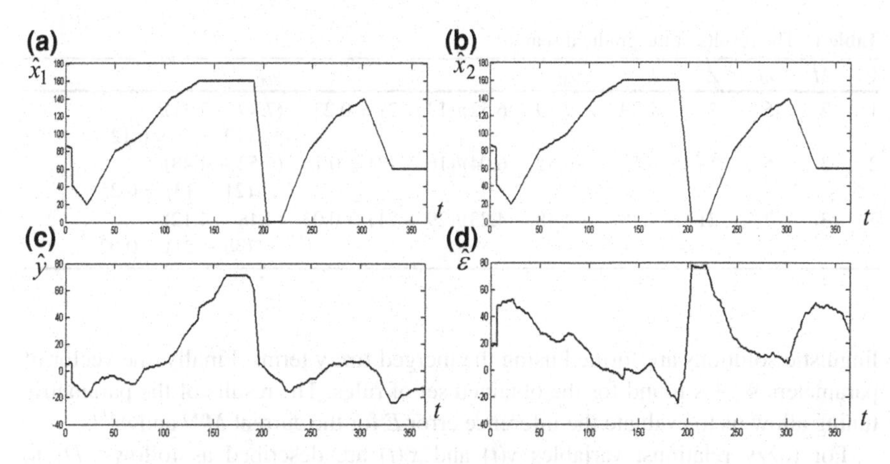

Fig. 1 Training data: **a** change of the demand for the produce; **b** stock quantity-on-hand change; **c** inventory action; **d** change of the produce remainder in store after control

5 Design of the Inventory Control System

Let us present the inventory control system in the form of the object $y(t) = f(x_1(t), x_2(t))$, where: $x_1(t)$ is demand at time moment t; $x_2(t)$ is stock quantity-on-hand at moment t; $y(t)$ is an inventory action at moment t, consisting in increasing–decreasing the stock. Training data for the district food-store house selling buckwheat is presented in Fig. 1a–c, where $t \in [1, 365]$ days; $x_1(t) \in [0, 200] \times 10^2$ kg; $x_2(t) \in [70, 170] \times 10^2$ kg; $y(t) \in [-100, 100] \times 10^2$ kg. The produce remainder in store after control $\varepsilon(t) = x_2(t) + y(t) - x_1(t)$ does not exceed the permissible inventory level, which is equal to 70×10^2 kg. The dynamics of the produce remainder $\varepsilon(t)$ change, presented in Fig. 1d is indicative of the control stability, i.e. of the tendency of index $\varepsilon(t)$ approaching a zero value.

The optimization problem can be formulated as follows.

Direct statement It is required to find M and m, for which $Z(M,m) \to$ min and $E(M,m) \leq \overline{E}$, where $\overline{E} = 7 \times 10^2$ kg is the permissible control error, i.e. the value of the «instantaneous» increase–decrease of the stock.

Dual statement It is required to find $M$and m, for which $E(M,m) \to$ min and $Z(M,m) \leq 21$.

The results of the gradient search are presented in Table 1.

Each iteration in Table 1 represents the results of the design for the current $M^{(k)}$ and $m^{(k)}$. Given the number of classes $M^{(k)}$, the vector of parameters $\mathbf{\Psi}_R^{(k)}$ is found for the predefined decision classes D_J. The obtained system of fuzzy relational equations (5) serves as the generator of rules for the current number of classes $m^{(k)}$. To define the number of rules Z and the support vector of input fuzzy terms $\mathbf{B}_a$, the solution set is generated for the predefined decision classes d_j. The obtained solutions can be translated into the fuzzy rules using the composite fuzzy terms for the significance measures $\mu^{c_{il}}(x_i)$ and $\mu^{D_J}(y)$. To reduce the number of terms q, the

Table 1 The results of the gradient search

k	M	m	Z	E	γ_M	γ_m
1	2	5	7	7.89	(7.89 – 6.52)/(13 – 7) = 0.23	(7.89 – 7.31)/ (12 – 7) = 0.12
2	3	5	13	6.52	(6.52 – 6.04)/(16 – 13) = 0.16	(6.52 – 4.48)/ (21 – 13) = 0.25
3	3	7	21	4.48	(4.48 – 4.23)/(29 – 21) = 0.03	(4.48 – 4.12)/ (38 – 21) = 0.02

linguistic solutions are formed using the merged fuzzy terms. Finally, the vector of parameters $\Psi_r^{(k)}$ is found for the obtained set of rules. The results of the parametric tuning allow us to evaluate the inference error E for the current $M^{(k)}$ and $m^{(k)}$.

For fuzzy relations, variables $y(t)$ and $x_i(t)$ are described as follows: D_1 to decrease the stock; D_2 stay inactive; D_3 to increase the stock; c_{i1} decreased (D); c_{i2} steady (St); c_{i3} increased (I).

The null variant of the fuzzy system is obtained for $M^{(1)} = 2, m^{(1)} = 5$.The system of fuzzy relational equations for $M = 2$ takes the form:

$$\mu^{D_1} = [(\mu^{c_{11}} \wedge 0.92) \vee (\mu^{c_{13}} \wedge 0.16)] \wedge [(\mu^{c_{21}} \wedge 0.25) \vee (\mu^{c_{23}} \wedge 0.89)]$$

$$\mu^{D_3} = [(\mu^{c_{11}} \wedge 0.28) \vee (\mu^{c_{13}} \wedge 0.95)] \wedge [(\mu^{c_{21}} \wedge 0.96) \vee (\mu^{c_{23}} \wedge 0.17)].$$

The significance measures $\boldsymbol{\mu}^D(d_1) = (0.88, 0.18)$, $\boldsymbol{\mu}^D(d_2) = (0.60, 0.24)$, $\boldsymbol{\mu}^D(d_3) = (0.36, 0.40)$, $\boldsymbol{\mu}^D(d_4) = (0.21, 0.59)$, $\boldsymbol{\mu}^D(d_5) = (0.17, 0.92)$ correspond to the classes: $d_1 = D_1$, sharply; $d_2 = D_1$, minimally; $d_3 = D_2$; $d_4 = D_3$, minimally; $d_5 = D_3$, sharply. Solution set of fuzzy relational equations for $m = 5$ is presented in Table 2, where variables $x_i(t)$ are described as follows: c_{i1}, sharply (sD); c_{i2} (St); c_{i3}, sharply (sI).

The solution of the direct problem is obtained for $M^{(2)} = 3, m^{(2)} = 5$.

The system of fuzzy relational equations for $M = 3$ takes the form:

$$\mu^{D_1} = [(\mu^{c_{11}} \wedge 0.96) \vee (\mu^{c_{12}} \wedge 0.65) \vee (\mu^{c_{13}} \wedge 0.16)] \wedge$$
$$[(\mu^{c_{21}} \wedge 0.16) \vee (\mu^{c_{22}} \wedge 0.80) \vee (\mu^{c_{23}} \wedge 0.98)]$$

$$\mu^{D_2} = [(\mu^{c_{11}} \wedge 0.83) \vee (\mu^{c_{12}} \wedge 0.72) \vee (\mu^{c_{13}} \wedge 0.53)] \wedge$$
$$[(\mu^{c_{21}} \wedge 0.90) \vee (\mu^{c_{22}} \wedge 0.72) \vee (\mu^{c_{23}} \wedge 0.48)]$$

$$\mu^{D_3} = [(\mu^{c_{11}} \wedge 0.15) \vee (\mu^{c_{12}} \wedge 0.94) \vee (\mu^{c_{13}} \wedge 0.99)] \wedge$$
$$[(\mu^{c_{21}} \wedge 0.95) \vee (\mu^{c_{22}} \wedge 0.65) \vee (\mu^{c_{23}} \wedge 0.15)]$$

The significance measures $\boldsymbol{\mu}^D(d_1) = (0.85, 0.22, 0.15)$; $\boldsymbol{\mu}^D(d_2) = (0.57, 0.40, 0.18)$; $\boldsymbol{\mu}^D(d_3) = (0.31, 1.00, 0.34)$; $\boldsymbol{\mu}^D(d_4) = (0.16, 0.35, 0.60)$; $\boldsymbol{\mu}^D(d_5) = (0.15, 0.25, 0.88)$ correspond to the classes $d_1 \div d_5$. Solution set of fuzzy relational equations for $m = 5$ is presented in Table 3, where variables $x_i(t)$ are described as follows: c_{i1}, sharply (sD); c_{i1}, minimally (mD); c_{i3}, minimally (mI); c_{i3}, sharply (sI).

Table 2 Solution set for the null variant

$x_1(t)$		$x_2(t)$		w	$y(t)$
$\mu^{c_{11}}$	$\mu^{c_{12}}$	$\mu^{c_{21}}$	$\mu^{c_{22}}$		
[0.88, 1], *sD*	[0, 1]	[0, 0.17]	[0.88, 1], *sI*	0.90	d_1
[0.60, 1], *sD*	[0, 0.28]	0.28	0.60, *St*	0.87	d_2
0.36	0.40, *St*	0.40, *St*	[0.36, 1], *sI*	0.96	
[0.36, 1], *sD*	0.40	[0.40, 1], *sD*	0.36	0.84	d_3
0.36	[0.40, 1], *sI*	0.40, *St*	[0.36, 1], *sI*	0.92	
0.25	0.59, *St*	[0.59, 1], *sD*	[0, 0.25]	0.80	d_4
0.17	[0.92, 1], *sI*	[0.92, 1], *sD*	[0, 1]	0.96	d_5

Table 3 Solution set for the direct problem

$x_1(t)$			$x_2(t)$			w	$y(t)$
$\mu^{c_{11}}$	$\mu^{c_{12}}$	$\mu^{c_{13}}$	$\mu^{c_{21}}$	$\mu^{c_{22}}$	$\mu^{c_{23}}$		
[0.85, 1], *sD*	[0, 0.15]	[0, 0.15]	[0, 0.15]	[0, 0.48], *mI*	[0.85, 1], *sI*	0.93	d_1
[0.85, 1], *sD*	0.48, *mD*	[0, 1]	[0, 0.15]	[0, 0.15]	[0.85, 1], *sI*	0.98	
[0.57, 1], *sD*	[0, 0.15]	[0, 0.15]	0.48, *mD*	0.48, *mD*	[0.57, 1], *sI*	0.92	d_2
[0, 0.15]	[0, 0.15]	0.57, *mI*	[0, 0.15]	[0, 0.15]	[0.57, 1], *sI*	0.85	
0.57, *mD*	0.57, *mD*	[0, 1]	[0, 0.15]	[0, 0.15]	0.57, *mI*	0.81	
0.65, *mD*	0.72, *mD*, *mI*	0.65, *mI*	0.65, *mD*	0.72, *mD*, *mI*	0.65, *mI*	0.88	d_3
[0.46, 1], *sD*	0.34	[0, 0.34]	[0.46, 1], *sD*	0.31	[0, 0.31]	0.85	
[0, 0.31]	0.31	[0.46, 1], *sI*	[0, 0.34]	0.34	[0.46, 1], *sI*	0.93	
0.53, *mD*	0.53, *mD*	[0.60, 1], *sI*	[0.60, 1], *sD*	[0, 0.16]	[0, 0.16]	0.90	d_4
[0, 0.16]	[0, 0.16]	0.60, *mI*	0.60, *mD*	0.60, *mD*	[0, 1]	0.87	
[0, 0.16]	[0, 0.16]	[0.60, 1], *sI*	[0, 0.15]	0.60, *mI*	0.60, *mI*	0.89	
[0, 0.16]	[0, 0.53], *mI*	[0.88, 1], *sI*	[0.88, 1], *sD*	[0, 0.16]	[0, 0.16]	0.95	d_5
[0, 0.16]	[0, 0.16]	[0.88, 1], *sI*	[0.88, 1], *sD*	0.53, *mD*	[0, 1]	0.99	

The solution of the dual problem is obtained for $M^{(3)} = 3, m^{(3)} = 7$. The significance measures $\mu^D(d_1) = (0.92, 0.14, 0.12); \mu^D(d_2) = (0.49, 0.35, 0.17); \mu^D(d_3) =$ (0.31,1.00,0.34); $\mu^D(d_4) = (0.16, 0.40, 0.50)$; $\mu^D(d_5) = (0.10, 0.14, 0.94)$ correspond to the classes $d_1 \div d_5$. The significance measures $\mu^D(d_{12}) = (0.75,$ $0.22, 0.15)$; $\mu^D(d_{45}) = (0.12, 0.18, 0.76)$ correspond to the intermediate classes: $d_{12} = D_1$, moderately; $d_{45} = D_3$, moderately. Solution set of fuzzy relational equations for $m = 7$ is presented in Table 4, where variables $x_i(t)$ are described as follows: c_{i1}, sharply (*sD*); c_{i1}, minimally (*mD*); c_{i2} (*St*); c_{i3}, minimally (*mI*); c_{i3}, sharply (*sI*).

The results of the parametric tuning of the obtained relations and rules are given in Fig. 2 and Table 5. Figure 3 depicts comparison of the model and the reference control and comparison of the produce remainder value in store after control.

Table 4 Solution set for the dual problem

$x_1(t)$			$x_2(t)$			w	$y(t)$
$\mu^{c_{11}}$	$\mu^{c_{12}}$	$\mu^{c_{13}}$	$\mu^{c_{21}}$	$\mu^{c_{22}}$	$\mu^{c_{23}}$		
[0.92, 1], *sD*	[0, 0.15]	[0, 0.15]	[0, 0.15]	[0, 0.48], *mI*	[0.92, 1], *sI*	0.98	d_1
[0.92, 1], *sD*	0.48, *mD*	[0, 1]	[0, 0.15]	[0, 0.15]	[0.92, 1], *sI*	0.96	
[0.75, 1], *sD*	[0, 0.15]	[0, 0.15]	[0, 0.15]	0.50, *St*	[0.75, 1], *sI*	0.92	d_{12}
0.65, *mD*	0.65, *St*	[0, 1]	[0, 0.15]	[0, 0.15]	[0.65, 1], *sI*	0.86	
0.65, *mD*	[0.65, 1], *St*	[0, 1]	[0, 0.15]	[0, 0.15]	0.65, *mI*	0.89	
[0.49, 1], *sD*	[0.49, 1], *St*	[0, 1]	[0, 0.15]	[0, 0.15]	0.49, *mI*	0.88	d_2
[0, 0.15]	0.49, *mI*	0.49, *mI*	[0, 0.15]	[0, 0.15]	[0.49, 1], *sI*	0.90	
0.49, *mD*	[0, 0.15]	[0, 0.15]	[0, 1]	[0.49, 1], *St*	[0.49, 1], *sI*	0.83	
[0.49, 1], *sD*	[0, 0.15]	[0, 0.15]	0.49, *mD*	0.49, *mD*	[0, 0.15]	0.91	
0.65, *mD*	[0.72, 1], *St*	0.65, *mI*	0.65, *mD*	[0.72, 1], *St*	0.65, *mI*	0.87	d_3
[0.46, 1], *sD*	0.34	[0, 0.34]	[0.46,1], *sD*	0.31	[0, 0.31]	0.86	
[0, 0.31]	0.31	[0.46, 1], *sI*	[0, 0.34]	0.34	[0.46, 1], *sI*	0.94	
[0, 0.16]	[0, 0.16]	0.50, *mI*	[0.50, 1], *sD*	[0.50, 1], *St*	[0, 1]	0.85	d_4
[0, 0.16]	[0, 0.16]	[0.50, 1], *sI*	[0, 0.15]	0.50, *mI*	0.50, *mI*	0.83	
[0, 1]	[0.50, 1], *St*	[0.50, 1], *sI*	0.50, *mD*	[0, 0.16]	[0, 0.16]	0.91	
0.50, *mD*	0.50, *mD*	[0, 0.15]	[0.50, 1], *sD*	[0, 0.16]	[0, 0.16]	0.96	
[0, 0.16]	0.54, *St*	[0.76, 1], *sI*	[0.76, 1], *sD*	[0, 0.16]	[0, 0.16]	0.85	d_{45}
[0, 0.16]	[0, 0.16]	0.65, *mI*	0.65, *mD*	[0.65, 1], *St*	[0, 1]	0.88	
[0, 0.16]	[0, 0.16]	[0.65, 1], *sI*	0.65, *mD*	0.65, *St*	[0, 1]	0.94	
[0, 0.16]	[0, 0.53], *mI*	[0.94, 1], *sI*	[0.94, 1], *sD*	[0, 0.16]	[0, 0.16]	0.95	d_5
[0, 0.16]	[0, 0.16]	[0.94, 1], *sI*	[0.94, 1], *sD*	0.53, *mD*	[0, 1]	0.99	

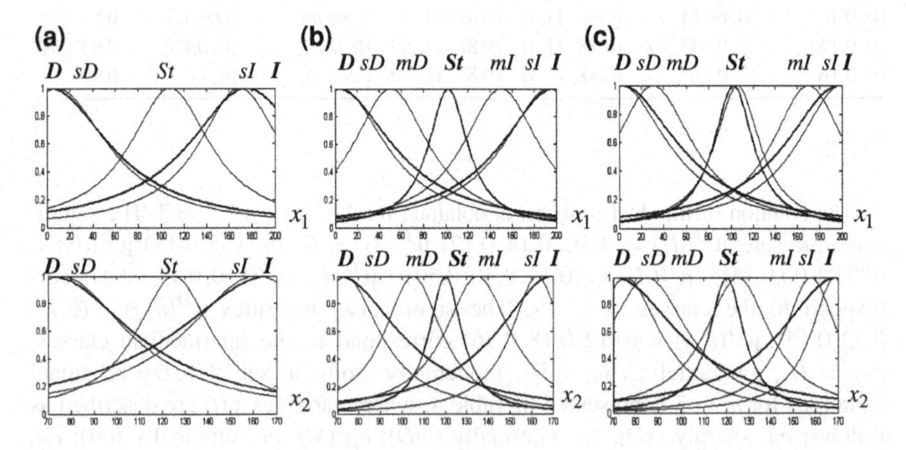

Fig. 2 Fuzzy terms membership functions for the null variant (**a**), the direct (**b**) and the dual (**c**) problem

Table 5 Membership functions parameters of variables x_1 and x_2 fuzzy terms

		$x_1(t)$					$x_2(t)$				
		D		*St*		*I*	*D*		*St*		*I*
Null variant	β	2.95				175.40	71.49				167.62
	σ	63.11				62.51	43.20				50.74
Direct problem	β	2.93		102.26		197.41	71.19		120.14		169.62
	σ	60.08		21.69		58.13	33.56		9.21		31.08
		sD	*mD*	*St*	*mI*	*sI*	*sD*	*mD*	*St*	*mI*	*sI*
Null variant	β	8.69		108.06		169.44	75.51		125.16		161.62
	σ	54.08		43.18		44.52	35.84		20.15		38.67
Direct problem	β	5.84	50.39		150.03	193.46	73.47	105.16		135.02	162.59
	σ	49.08	41.85		42.15	50.53	19.82	16.18		18.50	14.67
Dual problem	β	1.95	30.54	105.80	170.04	199.43	75.50	85.16	125.10	157.99	168.63
	σ	51.11	42.85	24.71	40.12	47.55	18.76	22.12	17.75	14.54	12.69

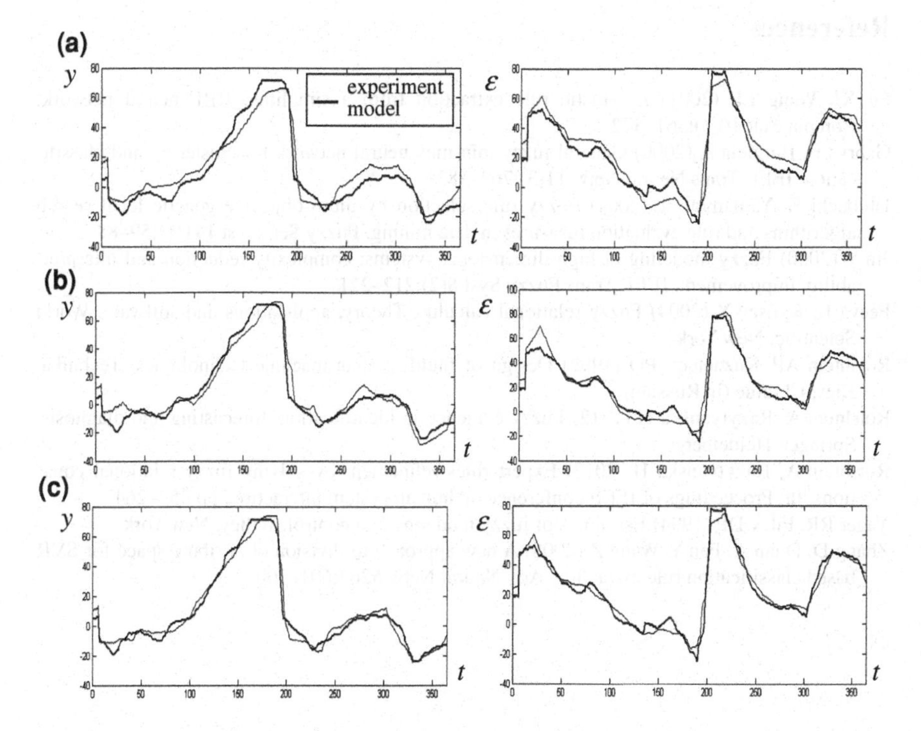

Fig. 3 Inventory action generated by the fuzzy model and produce remainder in store after control for the null variant (**a**), the direct problem (**b**) and the dual problem (**c**)

6 Conclusions

The innovative contribution of this paper is to propose an idea of building the composite knowledge base using fuzzy relations and rules. The system of fuzzy relations serves simultaneously as the support of the extracted regularities and as the generator of the rule-based solutions of fuzzy relational equations, where each solution represents a different trade-off between the number of fuzzy rules and the classification accuracy. The approach proposed can be extended to the case of three-objective genetic rule selection by including the optimization of the total rule length. For high-dimensional classification problems, the stage of rule generation should be augmented by the genetic local search algorithm for finding particular solutions of fuzzy relational equations. The local search procedure implies selection of the first level terms for each rule-based solution. Finally, the local search algorithm is combined with our approach to fuzzy rule selection.

References

Fu XJ, Wang LP (2001) Linguistic rule extraction from a simplified RBF neural network. Comput Stat 16(3):361–372

Gabrys B, Bargiela A (2000) General fuzzy min-max neural network for clustering and classification. IEEE Trans Neural Netw 11(3):769–783

Ishibuchi H, Yamamoto T (2004) Fuzzy rule selection by multi-objective genetic local search algorithms and rule evaluation measures in data mining. Fuzzy Sets Syst 141(1):59–88

Jin Y (2000) Fuzzy modeling of high-dimensional systems: complexity reduction and interpretability improvement. IEEE Trans Fuzzy Syst 8(2):212–221

Peeva K, Kyosev Y (2004) Fuzzy relational calculus. Theory, applications and software. World Scientific, New York

Rotshtein AP, Kuznetcov PD (1992) Design of faultless man-machine technologies. Technika, Kiev, Ukraine (in Russian)

Rotshtein A, Rakytyanska H (2012) Fuzzy evidence in identification, forecasting and diagnosis. Springer, Heidelberg

Rotshtein A, Rakytyanska H (2013) Expert rules refinement by solving fuzzy relational equations. In: Proceedings of IEEE conference on human system interaction, pp 257–264

Yager RR, Filev DP (1994) Essentials of fuzzy modeling and control. Willey, New York

Zhang D, Duan A, Fan Y, Wang Z (2008) A new approach to division of attribute space for SVR based classification rule extraction. Adv Neural Netw 5263:691–700

Author Index

Z. S. Hippe et al. (eds.), *Issues and Challenges in Artificial Intelligence*,
Studies in Computational Intelligence 559, DOI: 10.1007/978-3-319-06883-1,

Subject Index

Z. S. Hippe et al. (eds.), *Issues and Challenges in Artificial Intelligence*,
Studies in Computational Intelligence 559, DOI: 10.1007/978-3-319-06883-1,

Printed in the United States
By Bookmasters